TEUBNER-TEXTE zur Informatik   Band 24

T. Schnekenburger/G. Stellner (Hrsg.)

Dynamic Load Distribution for Parallel Applications

TEUBNER-TEXTE zur Informatik

Herausgegeben von
Prof. Dr. Johannes Buchmann, Darmstadt
Prof. Dr. Udo Lipeck, Hannover
Prof. Dr. Franz J. Rammig, Paderborn
Prof. Dr. Gerd Wechsung, Jena

Als relativ junge Wissenschaft lebt die Informatik ganz wesentlich von aktuellen Beiträgen. Viele Ideen und Konzepte werden in Originalarbeiten, Vorlesungsskripten und Konferenzberichten behandelt und sind damit nur einem eingeschränkten Leserkreis zugänglich. Lehrbücher stehen zwar zur Verfügung, können aber wegen der schnellen Entwicklung der Wissenschaft oft nicht den neuesten Stand wiedergeben.

Die Reihe „TEUBNER-TEXTE zur Informatik" soll ein Forum für Einzel- und Sammelbeiträge zu aktuellen Themen aus dem gesamten Bereich der Informatik sein. Gedacht ist dabei insbesondere an herausragende Dissertationen und Habilitationsschriften, spezielle Vorlesungsskripten sowie wissenschaftlich aufbereitete Abschlußberichte bedeutender Forschungsprojekte. Auf eine verständliche Darstellung der theoretischen Fundierung und der Perspektiven für Anwendungen wird besonderer Wert gelegt. Das Programm der Reihe reicht von klassischen Themen aus neuen Blickwinkeln bis hin zur Beschreibung neuartiger, noch nicht etablierter Verfahrensansätze. Dabei werden bewußt eine gewisse Vorläufigkeit und Unvollständigkeit der Stoffauswahl und Darstellung in Kauf genommen, weil so die Lebendigkeit und Originalität von Vorlesungen und Forschungsseminaren beibehalten und weitergehende Studien angeregt und erleichtert werden können.

TEUBNER-TEXTE erscheinen in deutscher oder englischer Sprache.

Dynamic Load Distribution for Parallel Applications

Edited by

Dr. Thomas Schnekenburger
Dr. Georg Stellner
Technische Universität München

Springer Fachmedien Wiesbaden GmbH
1997

Dr. Thomas Schnekenburger

Born 1964 in Villingen. Studies at the University of Stuttgart from 1984 to 1989. Received diploma in 1989 from the University of Stuttgart. Research assistant at the Technische Universität München from 1990 to 1997. Received Ph.D. in 1994 from the Technische Universität München. Project scientist at Siemens AG, Corporate Technology (ZT), Dept. Software beginning in 1997.

Dr. Georg Stellner

Born 1966 in Augsburg. Studied computer science at the Technische Universität München from 1986 to 1992. Received diploma in computer science in 1992. Researcher at the Technische Universität München from 1992 to 1997. Received Ph.D. from the Technische Universität München in 1996.
Research interests: Distributed and parallel programming environments, checkpointing and load balancing methods.

Gedruckt auf chlorfrei gebleichtem Papier.

Die Deutsche Bibliothek – CIP-Einheitsaufnahme

Schnekenburger, Thomas:
Dynamic load distribution for parallel applications /
Schnekenburger/Stellner. –
Stuttgart ; Leipzig : Teubner, 1997
   (Teubner-Texte zur Informatik ; Bd. 24)

   ISBN 978-3-8154-2309-7   ISBN 978-3-663-01522-2 (eBook)
   DOI 10.1007/978-3-663-01522-2

Umschlaggestaltung: E. Kretschmer, Leipzig

Preface

Dynamic load distribution for parallel applications is one of the most important topics of research in the area of parallel and distributed systems. The German Science Foundation has been supporting research in this area by funding SFB 342 "Methods and Tools for the efficient Use of Parallel Systems" and GK "Cooperation and Resource-Management in Distributed Systems" since 1990 and 1995 respectively. This manuscript is the result of research in both projects done jointly in the department for informatics and electrical engineering of Technische Universität München as well as by the industrial partner Corporate Research of Siemens AG, Munich.

First, surveys and classifications of existing models and implementations are given. Based on this comprehensive framework, several systems for checkpointing, a distributed thread kernel, load distribution in large workstation networks, and adaptive load distribution systems have been implemented and tested. The structure and results obtained with such systems are also covered. Finally a middleware - based architecture and three application areas are used to test the systems described beforehand.

Understanding the needs of large distributed applications and the efficient use of heterogenous networked resources is most important to make good use of the distributed computing power as it is available worldwide in the internet and also on a private basis in intranets. This manuscript will help the reader to understand the questions and can serve as a basis for future research and development as well as in training for distributed applications.

Thanks to the editors of Teubner Verlag for their support, for the editors Thomas Schnekenburger and Georg Stellner, to the authors of the different chapters and the many colleagues of SFB 342 and Graduiertenkolleg, that have contributed to this manuscript.

Munich, 24.06.1997   Arndt Bode
                                                                         Head of SFB 342

Contents

1 Introduction

Distributed computing environments such as loosely coupled parallel computers and workstation networks are more and more used as cost effective platforms for complex applications requiring large amounts of system resources. In contrast to tightly coupled parallel architectures using
common resources like main memory and external devices, parallelization in distributed architectures has to consider unequal distribution of load among resources and also the communication
amount between the components of the complete system.

This book emphasizes dynamic, application-oriented load distribution. That means, the goal of
load distribution is to dynamically assign workload of parallel applications to the distribution units
of the underlying system. Nevertheless, the main part of the subsequent classifications can also
be applied to system-oriented load distribution. According to the popular classification of CASA-
VANT and KUHL [1], the book deals with *dynamic global scheduling*. Following CASAVANT and
KUHL, global scheduling is to decide *where* to execute processes, whereas local scheduling is to
assign processes to time-slices of single processors. The topic of this book can be classified as
global since it deals with the decision of *where* to assign application workload. According to the
majority of work in the literature, we use the keyword *load distribution* instead of *global scheduling*. CASAVANT and KUHL classify load distribution as *dynamic*, if load distribution decisions are
made at runtime whereas *static* load distribution only uses a priori knowledge about the resource
needs. We regard *dynamic* load distribution for two reasons:

- For many parallel applications, the amount of workload of individual subproblems (for example the number of CPU cycles or the number of I/O-operations) cannot be predicted at
  startup-time (not to mention compile-time). Therefore, *dynamic* load distribution has to
  consider the actual resource requirements of the parallel application.
- Due to their excellent price-performance ratio, workstation clusters and networked PCs are
  more and more replacing expensive mainframe computers in the high-performance area. In
  contrast to mainframe computers that usually consist of homogeneous and exclusively used
  nodes, workstation clusters and networked PCs are often heterogeneous combinations of
  different machines that are used in multi-tasking mode. Therefore, the amount of available
  resources for individual processes of parallel applications is unpredictable. *Dynamic* load
  distribution has to consider this varying availability.

The most important difference between static and dynamic load distribution is that components
realizing dynamic load distribution consume resources and therefore imply some overhead. The
tradeoff between this overhead and the quality of load distribution is the main challenge of dynamic load distribution.

The next part presents surveys and classifications of load distribution. In contrast to [1] and
many other classifications (e.g. [3; 2]) and surveys of load distribution, we do not consider only
the strategy (Section 2.2), but present also a general classification (Section 2.1) with respect to the
objectives, integration level, structure and implementation of load distribution. Furthermore, there
are detailed classifications of load models (Section 2.3) and migration mechanisms (Section 2.4).
The proposed classification schemes are applied to several examples in Section 2.5. The classifications can be used to obtain a general view of load distribution. They can be used as a common

terminology for comparing different methods for load distribution. Furthermore, the classifications may be helpful to understand a given description of a load distribution system by classifying the system according to the proposed classification schemes.

Part 3 describes four systems supporting load distribution for parallel applications. Part 4 presents four systems with application integrated load distribution. The individual sections of Part 3 and 4 are ordered according to their integration level and generality. We start in Part 3 with general systems for load distribution that are not specific to an application domain and end in Part 4 with specific solutions that are directly integrated into applications. Section 3.1 describes *CoCheck*, a system for consistent checkpointing that can be used for transparent process migration of PVM applications. A distributed thread kernel with an integrated load distribution strategy is presented in Section 3.2. Section 3.3 describes *Dynasty*, an economic-based concept for load distribution in workstation networks. *ALDY*, a library supporting application-integrated load distribution, is shown in Section 3.4. Part 4 starts with the description of a middleware-based architecture for load distribution in Section 4.1. Two parallel applications realizing load distribution by dynamically assigning new subproblems to distribution units follow. Section 4.4 describes a parallel discrete event simulation based on the *time warp* concepts. All sections in Part 3 and 4 contain individual classifications according to the classification schemes proposed in part 2.

References

[1] T. L. CASAVANT and J. G. KUHL. A Taxonomy of Scheduling in General–Purpose Distributed Computing Systems. *IEEE Transactions on Software Engineering*, 14(2):141–154, 1988.

[2] T. LUDWIG. *Automatische Lastverwaltung für Parallelrechner*. Informatik 94. BI-Wissenschaftsverlag, 1993.

[3] L. TURCOTTE. A Survey of Software Environments for Exploiting Networked Computing Resources. Technical Report MSU-EIRS-93-2, NSF Engineering Research Center for Computational Field Simulation, Mississippi State University, Starkville, 1993.

2 Surveys and Classifications

Classification schemes are helpful tools in making the tasks of comparison and evaluation easier. Several such schemes for the purpose of classifying load distribution have been designed but so far none of them has been generally accepted. Furthermore, these schemes differ much in the extent to which technical details (like migration mechanisms etc.) have been included. This book, on the contrary, does not use one single classification scheme to cover all aspects of load distribution. Instead, several clear-cut and well-defined classification schemes for different, orthogonal aspects are presented.

Section 2.1 presents a general classification scheme regarding objectives, integration level, structure and implementation of load distribution The basic components of a system realizing load distribution, namely load distribution strategies, load models, and migration mechanisms are classified in Sections 2.2 to 2.4. Section 2.5 shows the application of the classification schemes to several examples.

2.1 General Classification of Load Distribution

by Thomas Schnekenburger

By definition, distributed systems consist of "distribution units" that provide resources for applications. Consequently, there arises the problem to assign certain objects, in the following called *entities*, to certain distribution units, in the following called *targets*. For example, entities may correspond to processes and targets may correspond to computational nodes. Generally, load distribution denotes the problem of managing the assignment of entities to targets.

This section discusses objectives of load distribution regarding the possible application domain, the intent of load distribution and the basic functions of load distribution. Possible integration levels for load distribution are shown. Furthermore, the problem structure of parallel applications is classified with respect to load distribution. Finally, possible realizations of entities and targets for load distribution are presented. The individual aspects are summarized to a general classification scheme.

2.1.1 Objectives

Independent from the realization of load distribution, a load distribution method can be classified according to basic objectives such as the application domain (i.e. is it specific to a certain domain), the intent (i.e. is it intended to be used as part of system management or as part of application management), and the function (i.e. is it used for the assignment of given entities or for partitioning).

The literature often mentions the keywords *load balancing* and *load sharing* instead of load distribution. Load balancing and load sharing can be defined as different specializations of load distribution. Load balancing usually denotes the objective to keep the load of targets (according to the selected load model) balanced. For example, a strategy may try to balance the number of processes in the CPU ready queue. Load sharing usually denotes the objective to share the resources of the system. In most cases this is accomplished by assigning entities to "idle" targets, or, from another point of view, to try to keep all targets busy. For example a central process scheduler may assign new jobs to idle nodes if idle nodes exist. We do not include these definitions into the general classification for two reasons: Firstly, it is often difficult to decide whether a given system tries to accomplish either load balancing or load sharing (using the definitions mentioned above). Secondly, there is a large number of papers that use the keywords load distribution, load balancing, and load sharing either interchangeably or by using an alternative definition. Therefore, this book uses always the keyword *load distribution* and assumes that load distribution covers load balancing as well as load sharing.

Domain: The first basic objective is the domain of a given load distribution method. A load distribution method is called *domain specific*, if it is designed for a specific application domain. For example, a load distribution method may be specifically designed for parallel database operators (e.g. [14]) or parallel image processing algorithms (e.g. [8]).

Intent: Regarding the intent of a load distribution method, there are two important areas for load distribution:

System-oriented: Large distributed computing environments are normally used by more than one application. Consequently, there is a need to distribute the components of applications so that the distributed resources of the system are efficiently used by applications.

Application-oriented: Parallel applications in distributed computing environments consist of several interacting processes that are spread over the computing units of the system. Load distribution for parallel applications tries to reduce the application's run time by exploiting the parallelism within the application and by assigning components of the parallel application so that the overhead due to synchronization and communication is kept small. There are two possible approaches to application-oriented load distribution:

> **Exclusive:** It is assumed that the application has exclusive access to the distributed system. In that case, load distribution either tries to assign the same amount of workload to each application process, assuming a homogeneous system or, in case of a heterogeneous system, it uses static information about resource capacities for assigning workload according to static "weights" of processes.
>
> **Adaptive:** The parallel application has to share resources with other applications. Resource usage of other applications cannot be explicitly controlled. Therefore, the parallel application has to *adapt* its workload distribution also to the actual resource usage of other applications.

Obviously, these aspects may be combined either by using an integrated approach or just by using application-oriented load distribution within an environment providing system-oriented load distribution.

Function: Generally, load distribution has to assign *entities* to *targets* in order to fulfill given requirements. Therefore, the primary function of any load distribution method is *assignment*: Entities (that may be given completely at program start or that may be generated dynamically) have to be assigned to certain targets. Assignment may be preemptive (*migration* of entities to other targets) or non-preemptive (*placement* of new entities at target).

Some load distribution methods provide *partitioning* as an additional functionality: In the partitioning phase, the concrete entities are formed. For example, a matrix is partitioned into several blocks or a database relation is partitioned into several hash files. Although the *principle* of partitioning is usually implemented by the application (for example, whether a matrix is partitioned

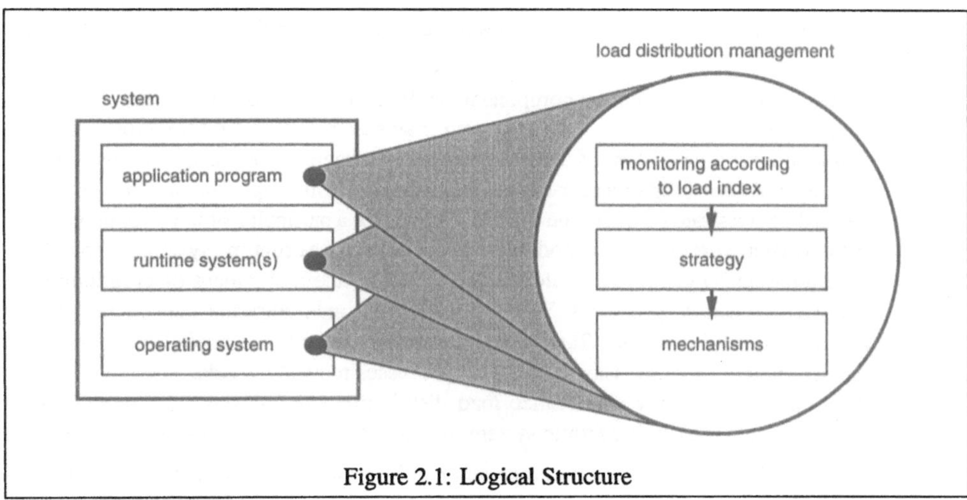

Figure 2.1: Logical Structure

one-dimensionally into "stripes" or two-dimensionally into rectangles), a load distribution method may manage the concrete partitioning (for example, determine the *number* of blocks or the *size* of hash files).

2.1.2 Integration Level

Logically, an environment using a load distribution system can be split into the managed system consisting of the application program, one or more runtime systems (for example, a communication subsystem or a programming language runtime system), and the operating system (cf. Figure 2.1).   Load distribution management is integrated into components of that system. Load distribution management logically consists of a monitoring component that is responsible for collecting information about the system that has to be managed. The monitoring component delivers actual information about the *load* of the system according to a particular *load index*. This information is used by the load distribution strategy to decide, whether to start one of the possible mechanisms for the assignment of entities to targets. That means, load distribution management is a part of the system consisting of monitoring according to a load index, strategy, and mechanisms.

According to Figure 2.1, we can identify three integration levels. Each of these levels may be used to integrate monitoring, strategy, or mechanisms, respectively.

Application: The load distribution component is realized within the application program.
- Regarding monitoring, the application can collect application related information about application objects. For example, a parallel database may collect statistical information about its data [14]. Time warp simulation programs monitor the virtual time of the simulation processes (4.4 on page 160).
- Regarding the strategy, the application may implement a strategy that considers the specific semantics of the application. For example, a searching algorithm may use problem specific heuristics to determine the order of scheduling subproblems.
- The application may implement specific mechanisms. For example, a simulation program may implement the migration of a complex data structure using knowledge about

implicit connections with other data structures of the simulation problem (4.4 on page 160).

Runtime systems: The load distribution component can be realized as part of a runtime system.

- Regarding monitoring, the runtime system may deliver internal information. For example, the runtime system of a communication package like PVM may deliver information about blocking of a process due to waiting for a message from another process.
- Regarding the strategy, a runtime system may realize a particular strategy for objects of the underlying programming model. As a special case, the runtime system of a parallel programming language may automatically manage the assignment of programming language objects to processes. For example, the Linda language constructs [4] are based on a "pool of tasks". Tasks are dynamically added to the pool. Linda manages the assignment of these tasks to "worker" processes requesting tasks.
- A runtime system may also realize load distribution mechanisms. As an example, consider again the Linda runtime system that assigns Linda tasks to worker processes.

Operating system: The load distribution component can be realized as part (or as an extension) of the operating system.

- Regarding monitoring, the operating system may deliver information about the states of operating system objects such as throughput numbers of resources and page faults of user processes.
- Regarding the strategy, the operating system may transparently assign operating system objects. For example, a distributed operating system may use the actual "load" of individual nodes to decide about the assignment of a new process.
- The operating system can realize load distribution mechanisms for operating system objects. For example, a distributed operating system may migrate user processes from overloaded to idle nodes.

The difference between the runtime system and the operating system level is that the runtime system is part of a normal application from the operating system's point of view. Obviously, the load distribution components need not to be integrated into a single level: The individual levels may be combined, leading to a large range of possible integration levels.

In some systems, the application program does not directly use underlying runtime systems for load distribution, but specific language constructs are used by a compiler to generate runtime system calls and additional information for the load distribution system. In that case, integration of monitoring, strategy and mechanisms may be classified as *compiler supported* (see for example section 3.2 on page 81 and [17]).

2.1.3 Structure

Regarding the structure of a parallel application, we can find two classes with respect to the potential for load distribution:

Location independency There is sufficient degree of freedom for assigning new subproblems of the parallel algorithm to the targets for load distribution. The corresponding programming paradigm realizes load distribution by selecting the target for a new subproblem according to the actual load of targets. In other words, load distribution for this class of applications

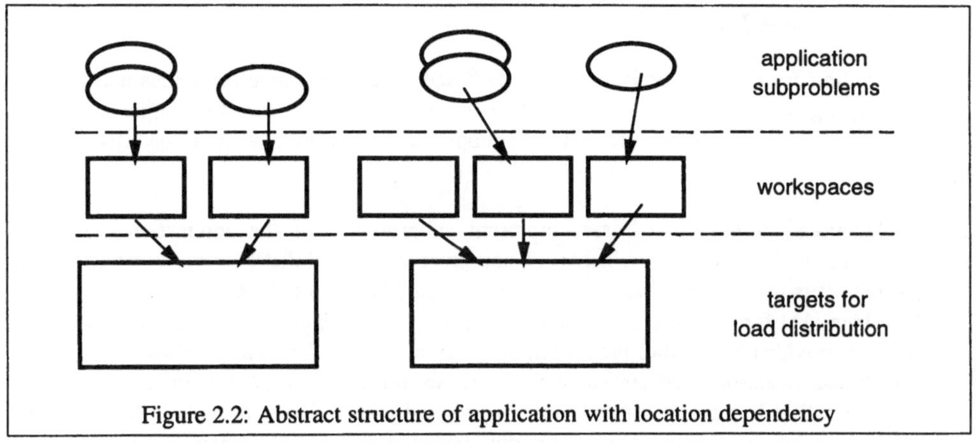

Figure 2.2: Abstract structure of application with location dependency

can be based on *placement* of subproblems. In the literature, this paradigm is called task-farming [18], task-bag [2], or self-scheduling.

Location dependency The degree of freedom for assigning subproblems to targets is too small to achieve sufficient efficiency. In that case, the general concept for realizing load distribution is to introduce an indirection: *Workspaces* (implicitly) represent locations that are used as targets for assigning application subproblems. Workspaces are assigned to the targets for load distribution. The corresponding programming paradigm can be modeled as a two-level mapping (Figure 2.2): Subproblems of the parallel algorithm are assigned to workspaces (for example a process). The mapping of subproblems to workspaces is restricted by the semantics of the application, that means, for a given subproblem, the semantics of the application determine which workspace receives the subproblem. Load distribution is realized by assigning and *migrating* workspaces to the targets for load distribution.

Backtracking algorithms such as traveling salesman are typical applications providing location independency. Nevertheless, there are many parallel algorithms as for example simulation programs and most numerical algorithms that belong to the location dependency class.

Note that location independency is not only a qualitative, but also a quantitative attribute. Consider a typical client-server structured application, where a client issues several queries to servers and each query may be handled by any server. If client requests are short queries that arrive frequently, the application is obviously location independent, because load distribution can be realized just by selecting a "lightly loaded" server for a new query. If client requests are not short queries but long running complex queries, the degree of freedom for assigning a query to a server remains the same. Nevertheless, even this degree of freedom may not be sufficient to achieve the desired performance: It may happen that when the system changes from a heavily loaded state to a less loaded state, several queries are still running on a certain server whereas the other servers are idle. In that case, some kind of migration of running queries may become necessary. Summing up, whether an application is location independent or not depends not only on the applications structure, but also on the quantitative attributes of resource requirements and on the desired performance of the system.

2.1.4 Entities and Targets

Load distribution for parallel applications can be based on a large range of entities and targets. This paragraph gives a survey of possible combinations. First we discuss the simple model for location independent applications consisting of subproblems that are placed on load distribution targets.

Process placement on nodes: Processes represent subproblems of the application. Process placement is the classical realization of non-preemptive load distribution by the operating system. It is used by distributed queueing systems like DQS [15], CODINE [6], and LSF [22] (see [11] for a survey of queueing systems). Furthermore, automatic process placement is provided by parallel programming environments as for example PVM [19]. Some distributed object-oriented environments as for example Mentat [7] provide *active objects* that are realized as processes and that are assigned automatically to nodes. As explained above, efficient load distribution that is only based on process placement requires location independent applications. The main problem of process placement is the large granularity of subproblems.

Placement of subproblems to processes: Subproblems of the application may be assigned to application processes, i.e., workload is distributed among the application processes. Subproblems may be represented as *threads* that are generated to execute a remote procedure call or just as messages that are received by processes and trigger some *computations*. The advantage compared to process placement is that there is a finer granularity of subproblems. On the other hand, load distribution that is only based on subproblem placement cannot dynamically add or delete nodes to the application. There are tools as for example Dynamo [20] that provide load distribution support for the placement of subproblems to processes.

As mentioned above, applications with location dependency generally assign subproblems not directly to nodes or processes, respectively, but to abstract *workspaces* that are assigned and migrated to the load distribution targets. There are the following possibilities for realizing workspaces:

Process migration to nodes: Processes represent *workspaces* that are assigned to nodes and that can be migrated between nodes. This corresponds to preemptive process assignment by operating systems. Although process migration is supported by several popular distributed queueing systems as for example CODINE [6], most of these systems do not support process migration in combination with process communication since internal information of other processes with respect to sockets etc. is not updated correctly. Recently some systems providing process migration mechanisms for processes using PVM were developed. The CoCheck system (section 3.1 on page 70) and the MPVM system [5] provide process migration for PVM applications. Regarding the general model in Figure 2.2 , subproblems correspond to *computation* that is performed by a process. The problem of process migration for load distribution of parallel applications is the significant overhead: Usually the entire memory has to be transferred to the new node, resulting in an increased network traffic. Furthermore, processes of parallel applications communicate and synchronize with the other processes of the application. Consequently, if a certain process is migrated, the other processes will also be slowed down due to the delayed interaction with that process.

Thread migration: In multithreaded environments, the system may assign and migrate threads to application processes, i.e., workload is distributed among the application processes. For example, the UPVM package [12] supports thread migration for PVM applications. If workspaces are realized as threads, subproblems correspond to computation that is performed by threads.

Application object migration: Using an object-based programming paradigm, workspaces may be realized as application objects that can be migrated to application processes. Subproblems of the application are realized as methods that are assigned to objects, i.e. a subproblem corresponds to an *object access*. Object migration is implemented by transferring the object's state to another process. Object migration is provided by some distributed object-oriented environments such as COOL [1]. Application object migration is also realized by several application integrated load distribution methods as for example [3; 10; 21]. [17] presents language constructs for integrating object migration into a parallel and distributed programming language.

Data migration: In parallel applications that work on a certain *domain* of data (such as a matrix or a mesh), workspaces may be represented by partitions of the data domain. Load distribution is realized by migrating (or re-distributing) the data domain among the processes of the application. [9] discusses several methods for data distribution in homogeneous environments. [8; 13; 16] describe methods for *adaptive* data distribution that takes into account heterogeneity and external load of the environment. The ADM system [5] provides support for data partitioning of PVM applications. In ADM a subproblem that is assigned to a workspace (data) corresponds to the computation on that data, i.e. a subproblem corresponds to a *data access*.

Migration of workspaces to load distribution targets may be combined with dynamic assignment of subproblems to workspaces. For example, migration of application objects to processes may be combined with the placement of processes to nodes.

2.1.5 Classification

Table 2.1 summarizes the general classification scheme. Regarding objectives, load distribution may be domain specific or not domain specific. The intent may be system-oriented, application exclusive assuming exclusive resource availability for the parallel application, application adaptive, or a combination of system-oriented and application-oriented load distribution. The function of load distribution is assignment, partitioning or both. Monitoring, strategy and realization of load distribution mechanisms may be realized within the application, within a runtime system, or within the operating system. Each of these components may be compiler-supported, i.e., interaction of the application and the component is realized by language constructs. The problem structure is either location independent or location dependent. Whereas location independent systems may use a simple model consisting of placement of subproblems (processes / threads / computation / object access / data access) to processes or nodes, location dependent systems require an intermediate level to represent a workspace (process / thread / object / data) that can be migrated. The difference between computation and object access / data access is that computation is directly associated with a process or thread that performs the computation whereas object access / data access is associated with an activity relating to a certain object or data region. Targets for distri-

Objectives	
domain specific	yes / no
intent	system-oriented / application exclusive / application adaptive / both
function	assignment / partitioning / both
Integration Level	
monitoring	application / runtime system / operating system
strategy	application / runtime system / operating system
mechanisms	application / runtime system / operating system
compiler support	none / monitoring / strategy / mechanisms
Structure	
problem structure	location independent / location dependent
subproblems	processes / threads / computation / object access / data access
workspaces	processes / threads / objects / data
targets	processes / nodes

Table 2.1: General Classification Scheme

bution (subproblem assignment for location independent systems and workspace assignment for location dependent systems) are either processes or nodes.

References

[1] P. AMARAL, C. JACQUEMOT, P. JENSEN, R. LEA and A. MIROWSKI. Transparent Object Migration in COOL-2. In *Proc. of Workshop on Dynamic Object Placement and Load-Balancing in Parallel and Distributed Systems, ECOOP'92, Utrecht, The Netherlands*, 1992.

[2] G. R. ANDREWS. Paradigms for Process Interaction in Distributed Programs. *ACM Computing Surveys*, 23(1):49–90, March 1991.

[3] C. BURDORF and J. MARTI. Load Balancing Strategies for Time Warp on Multi-User Workstations. *The Computer Journal*, 36(2):168–176, 1993.

[4] N. CARRIERO, D. GELERNTER, T. MATTSON and A. SHERMAN. The Linda alternative to message-passing systems. *Parallel Computing*, 20(4):633–656, 1994.

[5] J. CASAS, R. KONURU and S. W. OTTO. Adaptive Load Migration systems for PVM. In *Supercomputing'94, Washington D.C.*, pages 390–399, 1994.

[6] GENIAS Software GmbH. *Codine COmputing in DIstributed Networked Environments*, 1993.

[7] A. S. GRIMSHAW. Easy-to-use Object-Oriented Parallel Processing with Mentat. *Computer*, 26(5):39–51, 1993.

[8] M. HAMDI and C.-K. LEE. Dynamic Load Balancing of Data Parallel Applications on a Distributed Network. In *International Conference on Supercomputing*, Barcelona, Spain, 1995.

[9] R. V. HANXLEDEN and L. R. SCOTT. Load Balancing on Message Passing Architectures. *Journal of Parallel and Distributed Computing*, 13:312–324, 1991.

[10] K. A. HUA and H. C. YOUNG. A Cell–Based Data Partitioning Strategy for Efficient Load Balancing in A Distributed Memory Multicomputer Database System. Technical Report RJ 8041, IBM Research Report, 1991.

[11] J. A. KAPLAN and M. L. NELSON. A Comparison of Queueing, Cluster and Distributed Computing Systems. Technical Report Technical Memorandum 109025 (Revision 1), NASA Langley Research Center, Hampton, Virginia, 1994.

[12] R. KONURU, J. CASAS, S. OTTO, R. PROUTY and J. WALPOLE. A user-lvel process package for PVM. In *Scalable High-Performance Computing Conference*, pages 48–55. IEEE Computer Society Press, 1994.

[13] S. MIGUET and Y. ROBERT. Elastic load–balancing for image processing algorithms. In *First International Conference of the Austrian Center for Parallel Computation*, Salzburg, September 1991.

[14] E. RAHM and R. MAREK. Analysis of Dynamic Load Balancing Strategies for Parallel Shared Nothing Database Systems. In *Proceedings of the 19th International Conference on Very Large Data Bases, Dublin, Ireland*, 1993.

[15] L. REVOR. DQS users guide, 1992.

[16] T. SCHNEKENBURGER and M. HUBER. Heterogeneous Partitioning in a Workstation Network. In *8th Int. Parallel Processing Symposium, Workshop on Heterogeneous Computing, Cancun, Mexico*, pages 72–77. IEEE, 1994.

[17] T. SCHNEKENBURGER. Integration of Load Distribution into ParMod-C. Technical Report TUM-I9536 and SFB 342/19/95 A, Technische Universität München, 1995. http://wwwpaul.informatik.tu-muenchen/projekte/sfb342/pub/sfb342-19-95A.ps.gz.

[18] T. SCHNEKENBURGER. Efficiency of Dynamic Task Assignment in Multi-User Environments. In *11th Int. Conference on Systems Engineering, Las Vegas*, pages 408–413. University of Nevada, July 1996.

[19] V. SUNDERAM, G. GEIST, J. DONGARRA and R. MANCHEK. The PVM concurrent computing system: Evolution, experiences, and trends. *Parallel Computing*, 20(4):531–546, 1994.

[20] E. TÄRNVIK. Dynamo - A Portable Tool for Dynamic Load Balancing on Distributed Memory Multicomputers. *Concurrency: Practice and Experience*, 6(8):613–639, 1994.

[21] G. WEIKUM, P. ZABBACK and P. SCHEUERMANN. Dynamic File Allocation in Disk Arrays. In *Proceedings of the 1991 ACM SIGMOD International Conference on Management of Data*, pages 406–415, Denver, Colorado, 1991.

[22] S. ZHOU. LSF: load sharing in large-scale heterogeneous distributed systems. In *Workshop on Cluster Computing, Tallahassee, Florida*, December 1992.

2.2 Classification and Survey of Strategies

by Stefan Bischof and Thomas Erlebach

In this section we present a comprehensive classification scheme for load distribution strategies and give a brief survey of well-known and recently developed algorithms.

2.2.1 Classification of Load Distribution Strategies

Over the past 30 years, numerous strategies for load distribution have been proposed by researchers working in a large number of different areas of computer science. The individual strategies have sometimes been evaluated by thorough mathematical analysis, sometimes by simulation, and sometimes by appealing to the intuition of the reader. As a consequence, it is extremely difficult to compare different load distribution strategies and to determine which strategy will perform best in a particular scenario.

A fact that complicates this task even further is that most load distribution strategies proposed in the literature rely on very specific assumptions about the system and the application at hand. Unfortunately, such assumptions are often only implicit in the presentation of the strategy, and must be made explicit in order to determine whether a strategy is appropriate for a particular load distribution problem.

Our classification scheme for load distribution strategies is intended to make comparison and evaluation of different strategies easier. We note that the technical details of a particular implementation do not play a major role for the presentation and understanding of a strategy. Usually, a strategy need not specify for which particular kind of entities and targets it was designed. The same strategy could, in principle, be used to assign different kinds of entities (e.g., processes, threads, or database queries) to different kinds of targets (e.g., workstations, multiprocessor nodes, or data base servers). In addition, a strategy is not concerned with measuring load or calculating load

System Model	
Model Flavour	physical, combinatorial, microeconomic, random, fairness, none, ...
Target Topology	(heterogeneous) NOW, bus, mesh, fully connected, ...
Entity Topology	grid-like, tree-like, non-interacting entities, ...
Transfer Model	
Transfer Space	systemwide, long range, short range, neighborhood
Transfer Policy	preemptive, non-preemptive
Information Exchange	
Information Space	systemwide, long range, short range, neighborhood, central
Information Scope	partial, complete, none
Coordination	
Decision Structure	distributed, hierarchical, centralized
Decision Mode	autonomous, cooperative, competitive
Participation	global, partial
Algorithm	
Decision process	static, dynamic
Initiation	sender, receiver, sender & receiver, central, timer-based, threshold-based
Adaptivity	fixed, adjustable, learning
Cost Sensitivity	none, low, partial, high
Stability control	none, not required, partial, guaranteed

Table 2.2: Classification Scheme for Load Distribution Strategies

indices on the individual targets (see Section 2.3 for a classification of load models) or with the technical mechanisms used for assigning entities (see Section 2.4 for a classification of migration mechanisms). Furthermore, the local scheduling policy used for processing several entities on a single target is not considered part of the load distribution strategy. The classification scheme is intended to extract only the basic underlying strategy of an approach to a load distribution problem.

Of course, we do not want to imply that somebody looking for an adequate load distribution strategy should evaluate different strategies independently from a particular problem. Instead, it is essential to choose a strategy that matches the involved system and application requirements well. Therefore, the classification scheme was designed to include those aspects of a strategy that are necessary in order to estimate the expected performance on a particular system with certain properties.

The strategy classification scheme we propose is hierarchical and includes criteria pertaining to system model, transfer model, information exchange, coordination, and algorithm. The scheme is sketched in Table 2.2. A detailed description follows.

2.2.1.1 System Model

The underlying system model is one of the strongest classification criteria for load distribution strategies. If the system at hand does not match the system for which the strategy was designed, it is highly questionable whether the strategy can be of any use for that system. The relevant components of a system as seen by a load distribution strategy are the entities that represent the load and the targets to which the entities can be assigned. In addition, it is in general quite helpful to know where the motivation for the design of a particular strategy came from. Therefore, we include the following criteria:

Model Flavour: For several load distribution strategies it is obvious where the motivation for the selection of that particular strategy originated. Many strategies are based on the analogy with *physical* systems, e.g., the diffusion algorithm [7] or the gradient model algorithm [13]. Such strategies try to imitate physical systems where a kind of load distribution is accomplished as a consequence of the laws of nature.

Other strategies originate from *combinatorial* models (e.g., see [11] and 2.2.2.3). Here, the load distribution problem is usually formalized as a discrete mathematical problem and tackled by employing results from graph theory, scheduling theory or related fields.

Also economical systems have inspired researchers to design analogous load distribution strategies (e.g., see Section 3.3). This model is usually referred to as the *microeconomic* model.

If the target of a newly created or arriving entity is chosen at random, we classify the model as *probabilistic* or *random*. If a fair use of resources is the only obvious motivation for the strategy, the model flavour is *fairness*. (Note that our notion of *fair* should be understood in an intuitive sense, not related to any formal definitions of fairness that one can find in the context of distributed systems.) Some heuristics are not related to any particular analogy, and are therefore classified as model flavour *none*.

It should be mentioned that the above terms do not form a complete list, because strategies with a model flavour different from the ones mentioned here will surely be encountered now and then.

Target Topology: Parallel or distributed computer systems that are used in practice use a variety of different interconnection networks. There are bus-based shared memory multiprocessors, workstation clusters interconnected by ATM networks [18], or distributed memory multiprocessors with static (hypercube, torus) or dynamic (IBM SP2) network topologies [12].

Even though a strategy does not need to be restricted to any particular type of computer system, it still requires or assumes that the targets are interconnected in an appropriate way. This is due to the fact that a load distribution strategy must use the interconnection network to communicate load information and to transfer load objects.

In practice, distributed computer systems are often simply networks of workstations interconnected by Ethernet or ATM. Besides, modern parallel computer systems like IBM SP2 or Cray T3E (see [2]) have abandoned the classical store-and-forward routing in favour of more efficient routing paradigms, e.g., wormhole or virtual-cut-through routing or reconfigurable networks. With these developments, the communication latency does not depend much on the distance of the communication partners in the network anymore. Nevertheless, it is still necessary in such networks to limit the contention on individual links or buses. Hence, the target topology remains an important characteristic of parallel systems.

Another aspect of parallel systems that should be taken into account is whether they are homogeneous (all targets are the same type, and an executable file can be executed on any target) or heterogeneous. If a load distribution strategy can deal with heterogeneous systems, e.g., with heterogeneous NOWs (networks of workstations), the term *heterogeneous* is added to the entry for the criterion target topology in the classification scheme. Otherwise, it is assumed that the strategy deals only with homogeneous systems.

Entity Topology: Independent from the target topology, typical parallel application programs have a specific communication structure as well. For example, traditional solvers for differential equations distribute the data in a *grid-like* pattern and communicate intermediate results only among neighboring processes, whereas divide-and-conquer algorithms usually result in a *tree-like* communication pattern. Another possible scenario features *non-interacting entities* (abbreviated by *n.-i. entities*) that do not communicate with each other at all, e.g., sequential applications running on a parallel system.

Many load distribution strategies do not take the entity topology into account, i.e., they assume that the entities (load objects) are non-interacting. Obviously, such a strategy can cause severe problems if this assumption is not justified. It is likely to create high communication load by placing communicating entities at targets very distant from each other. Other strategies use information about the entity topology in order to keep communicating entities on neighboring targets.

2.2.1.2 Transfer Model

The basic task of a load distribution strategy is to transfer entities from heavily loaded targets to lightly loaded or idle targets. We refer to such transfers as *load transfers*. The following criteria specify which assumptions a strategy makes about the transfer model.

Transfer Space: Whereas some load distribution strategies transfer entities from heavily loaded targets only to neighboring targets, other strategies do not impose such a limit and transfer entities over greater distances in the network. In the former setting, the transfer space is *neighborhood*. If entities are transfered to non-neighboring targets, the transfer space can be *short range* (more than the direct neighbors, but still only a rather small part of the network), *long range* (a substantial part of the network, but not the whole network), or *systemwide* (no restrictions). The distinction between long range and short range can not be defined formally and remains intuitive in nature. In cases where such a distinction does not make sense, the term *restricted* can be used instead.

Transfer Policy: A crucial distinction between different load distribution strategies is whether they transfer an entity from one target to another even if the entity is already being processed (e.g., if the entity is a running UNIX process), or whether an entity is processed until completion on the target to which it has been assigned in the first place. This distinction between *preemptive* strategies employing migration of currently processed entities and *non-preemptive* strategies employing only initial mapping of entities to targets is ubiquitous in scheduling and load distribution theory.

2.2.1.3 Information Exchange

In addition to the communication overhead caused by the actual transfer of entities, a load distribution strategy increases the communication load of the network through the exchange of load-state information as well. It is desirable that this exchange of information uses up only a negligible amount of network resources, but it is also clear that good load distribution is difficult to achieve if the information available for evaluating the benefit of potential load transfers is outdated or incomplete.

Information Space: Whereas transfer space expresses the distance over which entities are transfered by the load distribution strategy, information space is concerned with the transfer of load-state information. Similar to the transfer space, *systemwide* information space means that the load distribution strategy transfers load-state information without restrictions on the distance of communicating targets in the network, and *restricted* information space can be further divided into *long range*, *short range* or *neighborhood*. Furthermore, there are strategies which transfer load-state information to a single central manager. In this case, the information space is *central*. Finally, strategies which do not transfer any load-state information have *empty* information space.

Information Scope: In addition to the information space, load distribution strategies also differ in the extent to which they collect load-state information before reaching a load-transfer decision. Bidding algorithms, for example, typically take into account only the load-state information from a small subset of all targets. Hence, they are classified as having *partial* information scope. The more recently developed precomputation-based load distribution algorithm, however, collects load-state information from the whole system (through local communications) in order to determine the actual load transfers (see [5] and 2.2.2.4). Therefore, its information scope is *complete*. As mentioned in 2.2.1.1, there are load distribution strategies that use no load-state information from other targets at all. The information scope of such strategies is classified as *none*.

2.2.1.4 Coordination

Typically, load distribution problems arise in distributed systems made up of a large number of more or less independent components. Hence, different load distribution strategies can also be characterized by how they make sure that load-transfer decisions can be reached effectively and in a coordinated manner in such a distributed system.

Decision Structure: Due to our limited intuition for distributed systems, load distribution strategies are understood most easily if they employ a central authority which gathers load-state information and makes all decisions regarding load distribution activities. Unfortunately, such a *centralized* decision structure is likely to create a bottleneck once the system grows larger.

At the opposite end of the spectrum, there are load distribution strategies where each target in the system can make decisions about potential load transfers. Such a *distributed* decision structure is scalable, but if there is too little coordination among the targets the effects of different load transfers may cancel each other out or even worsen the load imbalance.

A compromise between these two approaches is to employ a *hierarchical* decision structure. Such a strategy has many of the advantages of distributed decision making (including sufficient scalability) and avoids the bottleneck of a central decision maker by replacing it with a hierarchy of decision makers.

Decision Mode: If the load distribution strategy allows targets to decide about potential load transfers independently from each other, we classify the decision mode of the strategy as *autonomous*. For example, a strategy where an overloaded target transfers load to another randomly picked target has this property. Usually, however, targets cooperate in order to make sure that load transfers take place only if all involved targets agree. We call this latter alternative *cooperative* decision mode. A third possibility is *competitive* decision mode. Microeconomic strategies, for example, can usually be classified as competitive, because entities or targets compete for services mediated by brokers, without consideration for the needs of others.

Participation: This criterion classifies a load distribution strategy with respect to the number of targets which participate in load distribution activity at the same time. Many proposed load distribution strategies assume that all targets in a system jointly stop executing tasks at certain times and participate in a global load distribution phase until the load is sufficiently balanced. Frequently, it is also assumed that the system is synchronized and that all targets execute the load distribution activities in lock-step.

Whereas such strategies with *global* participation are often much easier to analyze, problems arise when they are to be applied in loosely coupled parallel computer systems where global synchronization is very time-consuming. Here, strategies with *partial* participation in load distribution activities seem much more appropriate. If the load on a certain subset of targets is imbalanced, only these targets participate in load transfers while the remaining targets can continue to execute their normal task load.

2.2.1.5 Algorithm

The classification criteria subsumed here are intended to capture general characteristics of the load distribution algorithm used to implement the strategy.

Decision Process: The decision process of a load distribution algorithm is *static* if it does not depend on load-state information accumulated at run-time. Static load distribution algorithms include compile-time partitioning of parallel applications and mapping of tasks to predefined or random locations.

Typically, however, the term "load distribution algorithm" already implies a *dynamic* decision process, whereas static load distribution algorithms are more commonly referred to as partitioning algorithms, mapping algorithms, or embedding algorithms. This classification scheme is intended to be used for dynamic load distribution algorithms, and hence many of the other criteria do not apply to static algorithms.

Initiation: The detection of load-imbalance in the system is a problem that must necessarily be addressed when one needs to employ a load distribution strategy in practice. Possible alternatives for strategies with partial participation are *sender* initiation with load distribution activity being initiated by overloaded targets, and *receiver* initiation with load distribution activity being initiated by underloaded or idle targets. These two can also be combined (*sender & receiver* initiation). Another possibility is to have a *central* component that initiates load distribution activity. In addition, it is possible to differentiate between *timer-based* and *threshold-based* initiation. The former refers to a strategy that performs load distribution activity after fixed time intervals, the latter to a strategy that starts load distribution as soon as load or load imbalance exceeds a certain threshold.

Adaptivity: An important aspect concerning the flexibility of a load distribution strategy is its adaptivity. A *fixed* strategy is independent of the current overall load level and of the characteristics of the entities present in the system. Since the required load distribution activity depends heavily on these dynamically changing factors, however, many researchers have designed strategies which are *adjustable* to the current load or other properties of the system. Another interesting approach are *learning* strategies that try to improve the effectiveness of their load distribution activities by learning from past experiences.

Cost Sensitivity: A load distribution algorithm should not disregard the costs that are necessarily incurred by every kind of load distribution activity, i.e., communication overhead and migration costs. After all, load distribution is usually only a means to achieve the goal of short response-times or high through-put of a system. If a load distribution algorithm keeps the load in the system perfectly balanced at all times but slows down the overall system substantially by the overhead created, this is not a satisfactory solution. Therefore, algorithms should also be classified according to the extent to which they take the costs imposed by load distribution activities into account. The cost sensitivity of a load distribution algorithm is classified as *none, low, partial* or *high.*

Stability Control: One of the shortcomings of certain load distribution strategies is that they may lead to system instability. A load distribution algorithm is part of the overall system, and the gathering of load-information and the transfers of entities form one target to another contribute to the total load. Furthermore, the load situation may change substantially from time to time, and it is important that the load distribution algorithm does not increase the system load inadequately in such moments due to useless load transfers. In the worst case, it may happen that all resources of the system are occupied by load distribution activities, without any progress being made towards a more balanced state or a more lightly loaded system.

Therefore, a load distribution algorithm should be designed to keep the system stable at any time. For some algorithms, stability control (i.e., particular steps taken in order to avoid system instability) is *not required,* because any system instability is made impossible by the inherent nature of the algorithm. Stability is *guaranteed* if the load distribution algorithm employs special means that keep the system stable under all conditions. For example, an algorithm can restrict the number of migrations of a single entity. *Partial* stability control adjusts certain parameters according to the current load-situation or according to the number of transfers of entities that happened in the recent past. But in the worst-case these precautions might fail to keep the system stable. Finally, there are algorithms that do not take instability problems into account at all. These are classified as having stability control *none.*

Note that the individual classification criteria are not completely orthogonal to each other. In the following, we summarize the obvious dependencies. Classifying transfer space or information space as neighborhood, short range or long range makes sense only if the target topology is not bus-based or fully connected. If information scope is none, information space must be empty.

If decision making is centralized, decision agreement must be classified as cooperative (viewing the fact that all targets agree to leave the load distribution decisions to the central authority as a kind of passive cooperation). If the decision process is static, the criteria pertaining to information exchange and to coordination, as well as the criteria initiation, adaptivity and stability control do not seem to be applicable.

Finally, it should be noted that we have decided not to include the specific criterion which a strategy tries to optimize (e.g., system throughput or total execution time) in the classification scheme. The reason is that the exact optimization criterion is not even specified for many strategies. Furthermore, given a specific optimization criterion and complete advance knowledge about all application characteristics (task execution times, task dependencies, communication requirements), the optimization problem is $\mathcal{NP}$-hard in almost every scenario [9]. Hence, load distribution strategies are usually heuristic approaches, and the quality of a strategy with respect to a certain criterion can frequently only be estimated by simulations.

2.2.2 A Survey of Load Distribution Strategies in Examples

Due to the vast amount of papers written about load distribution strategies, a comprehensive survey would be beyond the scope of this section. So we restrict this survey to four typical examples in order to reveal basic techniques for load distribution. The *diffusion approach* and the *bidding algorithm* are "classical" strategies that are used in quite a number of systems and applications. In addition, we describe two novel algorithms with interesting properties. A number of other strategies are surveyed in [19; 20].

2.2.2.1 The Diffusion Approach

Diffusion is a well-known load distribution strategy (LDS for short) and has been evaluated theoretically [7] as well as in practical experiments [19]. The diffusion LDS works like this: at the beginning of a load-distribution round every target collects the load values of all of its neighbors. Consider two neighboring targets i and j. If the load of i is greater than the load of j then i sends a certain part of this load-difference to j. Otherwise i receives some load from j. The amount of load that is transferred depends on the structure of the network. At the end of every load distribution round the targets update their load values. The diffusion LDS performs a certain number of rounds necessary to achieve a nearly balanced state.

Formally, if we denote the load distribution of n targets at the beginning of round $t \geq 1$ by $(l_1^t, l_2^t, \ldots, l_n^t)$ the load exchange is given by the equation

$$l_i^{t+1} = l_i^t + \sum_{j \in \Gamma(i)} \alpha_{ij}(l_j^t - l_i^t), \quad 1 \leq i \leq n,$$

where $\Gamma(i)$ is the set of direct neighbors of i and $0 < \alpha_{ij} < 1$ is the diffusion parameter of i and j. As a target can't give away more load than it possesses, we must have $1 - \sum_{j \in \Gamma(i)} \alpha_{ij} \geq 0$ for $1 \leq i \leq n$. CYBENKO [7] showed that this iteration converges against the uniform distribution if certain conditions hold.

Of course, the question arises how to set the diffusion parameters to obtain the fastest convergence. For example, it can be shown for mesh connected networks of dimension $d \geq 1$ that $\alpha_{ij} \equiv 1/(2d)$ is optimal [21]. This means that the load of i (not on the border of the mesh) is just the average of the load of its neighbors after each round.

Figure 2.3 shows one iteration of the diffusion algorithm in 4×4 mesh using the optimal diffusion

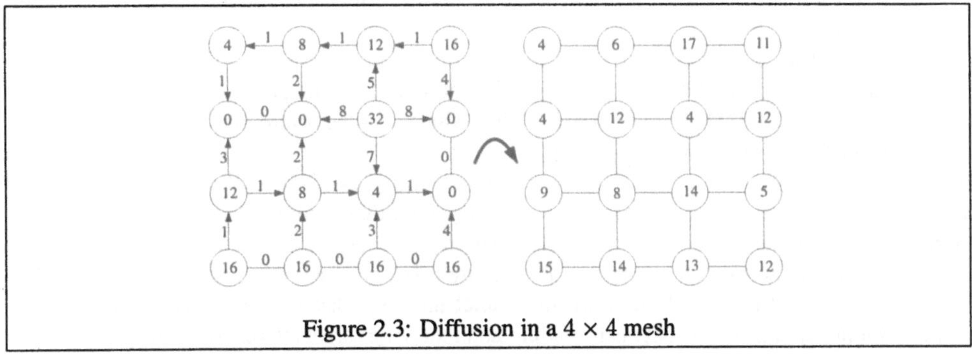

Figure 2.3: Diffusion in a 4×4 mesh

parameter $\frac{1}{4}$. Each target is labeled by its load value and the arrows indicate the amount and direction of the load transfer. Note that the average load is 10 and that the maximal deviation from the mean is reduced from 22 to 7.

In practice, the assumptions that all targets perform load distribution synchronously and the load is arbitrarily divisible are unrealistic, but it is possible to adapt the diffusion LDS for asynchronous systems and integer load values [19; 16].

2.2.2.2 Bidding and Balanced Allocations

One common procedure to obtain a fair price for a merchandise is to put it up for auction. Let us briefly describe a simple, no-minimum, one-round auction:

An auctioneer starts with a request for bids for a certain object to a set of bidders and then waits for bids. If any bids are made within a given period of time, the auctioneer selects the best one and knocks down the object to the highest bidder (ties are broken arbitrarily by the auctioneer).

This is the basic idea of bidding algorithms. To perform load distribution, there are two natural possibilities for the job of the auctioneer:

- **Overloaded targets** select some load for transfer to other, hopefully less loaded targets and send requests for bids for the selected entities to a certain number of targets (e.g., its direct neighbors). The targets evaluate the offered piece of load with regard to their own load situation and may choose to return a bid containing their rating of the entity or their current load index. If the initiating target receives any suitable bid in a certain amount of time and is still overloaded, it transfers the entity to the chosen bidder. Otherwise another request for bids is sent to a greater or different set of potential bidders (see [17] for a detailed description of this strategy).

- **Underloaded targets** offer their unused capacity instead of burdensome load. This strategy was proposed by [14] as a distributed "drafting" algorithm for load distribution.

If a target is allowed to make several bids in different auctions that take place concurrently, the strategy has to make sure that all transfers can be fulfilled and that the system remains stable. See [15] for implementation details of the bidding strategy.

Since the analysis of the bidding strategy in a distributed system is very difficult, we now consider an allocation policy [4] that is similar to the bidding strategy and admits a thorough analysis. Suppose that we place n balls (entities) one-by-one into n bins (targets) by putting each ball into a randomly chosen bin. It can be shown that with high probability there are $\Theta(\ln n/\ln\ln n)$ balls in the fullest bin after placing all n balls.

If we choose instead $d \geq 2$ bins independently and uniformly at random for every ball and put it into the least full among these bins, then [4] showed that the number of balls in the fullest bin drops to $\ln\ln n/\ln d + \Theta(1)$ with high probability when $n \to \infty$. Note that this strategy results in an exponential decrease of the maximum load.

The selection of d bins is directly related to the request for bids of the auctioneer of the bidding strategy, and the chosen bins make bids by responding with their current load. Of course, there is a trade-off between the number of bins queried and the resulting maximum load. The problem of finding optimal values for d when placement costs are taken into account is studied in [8].

This policy can be extended [1] to work in parallel and asynchronous systems by introducing several communication rounds between sending and receiving targets to resolve conflicts.

2.2.2.3 Load Distribution by Random Matchings

Many parallel and distributed systems can be modeled as a graph $G = (V, E)$, where the nodes of the graph represent targets and the edges represent direct communication links between targets. A matching of G is any subset of edges $M \subseteq E$ such that no two edges of M share an endpoint (see Figure 2.4).

Matchings are useful to perform local load distribution in parallel: if a node i is an endpoint of an

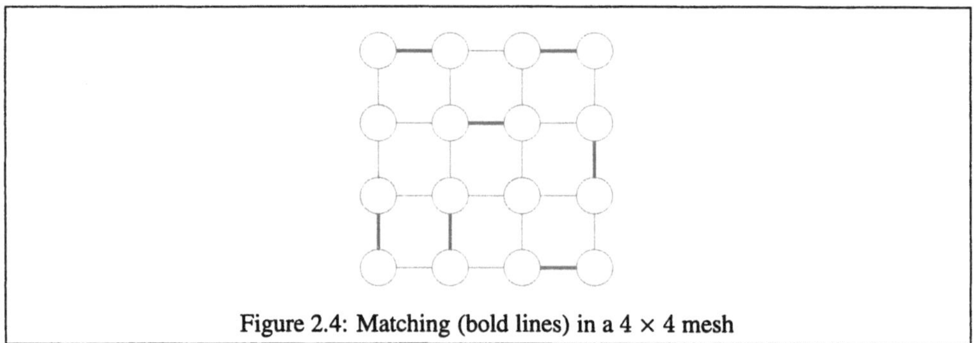

Figure 2.4: Matching (bold lines) in a 4×4 mesh

edge $e = \{i, j\}$ in the matching, it tries to reduce the load difference with the other endpoint j. Since no two edges of a matching are adjacent, any node is involved in at most one load transfer. This is a desirable feature due to limited communication capabilities of network interfaces.

The random matchings LDS [11] performs a certain number of synchronous rounds of the following form:

1. Generate a random matching M of G by a distributed algorithm.
2. Equalize the load of the endpoints of each edge in M.

The exact number of rounds necessary to achieve a nearly balanced load distribution with high probability depends on the degree of connectivity of the system.

To describe step 1 more precisely, let d be the maximum number of neighbors of any node i and

S_i a set of edges maintained by every node i. All S_i are empty at the beginning of each round. The generation of a random matching proceeds in two steps executed in parallel by every node:

(a) For every incident edge e: insert e into S_i with probability $\frac{1}{8d}$.

(b) Resolve conflicts by executing an agreement protocol. This protocol makes sure that
   - there is at most one edge in S_i and
   - if $S_i = \{\{i,j\}\}$ then $S_j = \{\{j,i\}\}$ (and consequently the union of all S_i, $i \in V$, constitutes a matching of G).

An important characteristic of a random matching generated by the above algorithm is that the probability to be part of the matching is at least $1/8d$ for each edge.

To evaluate the effectiveness of an LDS we need a measure for the imbalance in the system. The most common measure is the difference between the highest and lowest load of any target. In large systems however, this criterion might not allow to judge the overall load situation correctly. Therefore, it is reasonable to define the potential of a load distribution $l_1, l_2, \ldots, l_n$ in a system of n targets as $\Phi = \sum_{i=1}^{n} (l_i - \bar{l})^2$, where $\bar{l}$ denotes the average load. Clearly, the influence of a single target is reduced considerably according to this measure (see [3] for a detailed discussion of a wider class of imbalance measures).

It can be shown [11] that each round of the random matching LDS reduces the potential by a factor depending on the system topology if the potential is sufficiently large (otherwise load distribution isn't necessary anyway). Furthermore, step 2 of the algorithm can be modified so that at most one unit of load is transferred in each round. This variant is analyzed in [10].

2.2.2.4 Precomputation-Based Load Distribution

All LDSs presented so far reach their load transfer decisions on a small information base. Since most load transfers are very time-consuming (especially when they are preemptive), it is only natural to study algorithms that gather more load information in order to avoid superfluous load transfers.

One way of reducing load transfers is to compute a load distribution scheme before any transfers take place. Let $G = (V, E)$ be a graph representing a parallel or distributed system. We replace each undirected edge $e = \{u, v\}$ in E by two directed edges (u, v), (v, u) to indicate the direction of the load transfers. A load distribution scheme is a function δ that assigns a transfer value to each directed edge such that:

1. $\delta((u, v)) = -\delta((v, u)) \quad \forall \{u, v\} \in E,$
2. $l_v + \sum_{(u,v)} \delta((u, v)) = \bar{l} \quad \forall v \in V,$

where l_v is the load of node v and $\bar{l}$ denotes the average load. For simplicity, we assume that $\bar{l}$ and the δ-values are integers. Load distribution is performed by moving $\delta((u, v))$ load units from u to v if this value is positive. If $\delta((u, v)) < 0$ then u receives this much load from v. Hence, a load distribution scheme is like a road map that we can use to distribute the system load equally.

The precomputation based LDS presented in [5] (c.f. also [6, p. 23–95]) is designed for tree-connected systems. The reason is that a load distribution scheme can be computed efficiently for trees. The algorithm consists of two phases:

1. Precompute a load distribution scheme. This requires three steps:

   (a) Starting from the leaves of the tree, every node v (with exception of the root r) calculates the total load of the subtree rooted at v and sends this value to its parent.

	Diffusion	Bidding	Matchings	Precomputation
System Model				
Model Flavour	physical	microeconomic	combinatorial	combinatorial
Target Topology	mesh	fully connected	arbitrary	tree
Entity Topology	n.-i. entities	n.-i. entities	n.-i. entities	n.-i. entities
Transfer Model				
Transfer Space	neighborhood	systemwide	neighborhood	neighborhood
Transfer Policy	preemptive	non-preemptive	preemptive	preemptive
Information Exchange				
Information Space	neighborhood	systemwide	neighborhood	neighborhood
Information Scope	partial	partial	partial	complete
Coordination				
Decision Structure	distributed	distributed	distributed	hierarchical
Decision Mode	cooperative	competitive	cooperative	cooperative
Participation	global	partial	global	global
Algorithm				
Decision process	dynamic	dynamic	dynamic	dynamic
Initiation	timer-based	sender & receiver	timer-based	central
Adaptivity	fixed	fixed	fixed	fixed
Cost Sensitivity	none	none	none	none
Stability control	none	none	none	none

Table 2.3: Classification of Strategy Examples

(b) The root calculates the average load $\bar{l}$ and broadcasts $\bar{l}$ to all nodes.

(c) After receiving the broadcast-message, every node $v \in V \setminus \{r\}$ computes $\delta(v, parent(v))$. This is accomplished by subtracting $\bar{l} * \#nodes\_of\_subtree$ from the total load of the subtree rooted at v. Finally, v sends this value to its parent.

2. Perform the actual load transfers according to the load distribution scheme. This is done in a number of rounds because a node might have to wait for some of the load it must transfer. The number of rounds, however, is bounded by the diameter of the tree.

It is straightforward to generalize this algorithm to system-topologies that are cross products of trees. A grid, for example, is the cross product of two linear arrays. First, the algorithm balances all rows in parallel and is then applied a second time to balance the columns of the grid.

In Table 2.3, the four load distribution strategies discussed in this section are classified according to our classification scheme. It should be mentioned that the initiation of load distribution activities is not clearly specified for the diffusion strategy [7], for the strategy that employs random matchings [11], and for the precomputation-based strategy [5]. These strategies can also be implemented using different initiation than given in the table. Furthermore, we'd like to note that stability control may be considered as *not required* for these three strategies, because they employ load distribution phases with global participation. These load distribution phases always result in a balanced system. Nevertheless, it is possible that load distribution phases are initiated at unfavorable moments and cause an unnecessary slowdown of the system, and hence we choose to classify stability control for these strategies as *none*.

References

[1] M. ADLER, S. CHAKRABARTI, M. MITZENBACHER and L. RASMUSSEN. Parallel Randomized Load Balancing. In *27th Annual ACM Symp. on Theory of Computing STOC '95 (Las Vegas, Nevada, USA, May 29 – June 1, 1995)*, pages 238–248, New York, 1995. ACM SIGACT, ACM Press.

[2] G. S. ALMASI and A. GOTTLIEB. *Highly Parallel Computing*, volume 2. The Benjamin/Cummings Pubishing Company, Inc., Redwood City, CA, 1994.

[3] B. AWERBUCH, Y. AZAR, E. F. GROVE, M.-Y. KAO, P. KRISHNAN and J. S. VITTER. Load Balancing in the L_p Norm. In *36th Annual Symp. on Foundations of Computer Science (Milwaukee, Wisconsin, October 23–25, 1995)*, pages 383–391, Los Alamitos-Washington-Brussels-Tokyo, 1995. IEEE Computer Society, IEEE Computer Society Press.

[4] Y. AZAR, A. BRODER, A. KARLIN and E. UPFAL. Balanced Allocations. In *26th Annual ACM Symp. on Theory of Computing STOC '94 (Montreal, Quebec, Canada)*, pages 593–602, New York, 1994. ACM SIGACT, ACM Press.

[5] M. BÖHM and E. SPECKENMEYER. Precomputation Based Load Balancing. In *4th Workshop on Parallel Systems and Algorithms PASA '96 (Jülich, Germany, April 10–12, 1996)*, pages 173–190, Singapore, 1996. World Scientific Publishing Co.

[6] M. BÖHM. *Verteilte Lösung harter Probleme: Schneller Lastausgleich.* Ph.d. thesis, Mathematisch-Naturwissenschaftliche Fakultät, Universität zu Köln, 1996.

[7] G. CYBENKO. Dynamic load balancing for distributed memory multiprocessors. *J. Parallel Distrib. Comput.*, 7:279–301, 1989.

[8] T. DECKER, R. DIEKMANN, R. LÜLING and B. MONIEN. Towards Developing Universal Dynamic Mapping Algorithms. In *7th IEEE Symp. on Parallel and Distributed Processing 1995 (San Antonio, Texas, October 1995)*, pages 456–459, Los Alamitos-Washington-Brussels-Tokyo, 1995. IEEE Computer Society, IEEE Computer Society Press.

[9] M. R. GAREY and D. S. JOHNSON. *Computers and intractability. A guide to the theory of $\mathcal{NP}$-completeness.* W.H. Freeman and Company, New York-San Francisco, 1979.

[10] B. GHOSH, F. LEIGHTON, B. MAGGS, S. MUTHUKRISHNAN, C. PLAXTON, R. RAJARAMAN, A. RICHA, R. TARJAN and D. ZUCKERMAN. Tight Analyses of Two Local Load Balancing Algorithms. In *27th Annual ACM Symp. on Theory of Computing STOC '95 (Las Vegas, Nevada, USA, May 29 – June 1, 1995)*, pages 548–558, New York, 1995. ACM SIGACT, ACM Press.

[11] B. GHOSH and S. MUTHUKRISHNAN. Dynamic Load Balancing in Parallel and Distributed Networks by Random Matchings. In *6th Annual ACM Symp. on Parallel Algorithms and Architectures, SPAA'94 (Cape May, New Jersey, June 27 – 29, 1994)*, pages 226–235, New York, 1994. ACM SIGACT, ACM SIGARCH, ACM Press.

[12] F. T. LEIGHTON. *Introduction to parallel algorithms and architectures: Arrays, trees, hypercubes.* Morgan Kaufmann Publishers Inc., San Mateo, CA, 1992.

[13] F. C. LIN and R. M. KELLER. The Gradient Model Load Balancing Model. *IEEE Trans. Softw. Eng.*, SE-13(1):32–38, January 1987.

[14] L. M. NI, C.-W. XU and T. B. GENDREAU. A Distributed Drafting Algorithm for Load Balancing. *IEEE Trans. Softw. Eng.*, SE-11(10):1153–1161, 1985.

[15] R. RADERMACHER. *Eine Ausführungsumgebung mit integrierter Lastverteilung für verteilte und parallele Systeme.* Ph.d. thesis, Fakultät für Informatik der Technischen Universität München, 1996.

[16] J. SONG. A partially asynchronous and iterative algorithm for distributed load balancing. *Parallel Computing*, 20:853–868, 1994.

[17] J. A. STANKOVIC and I. S. SIDHU. An Adaptive Bidding Algorithm For Processes, Clusters and Distributed Groups. In *4th Int. Conf. on Distributed Computing Systems*, pages 49–59. IEEE Computer Society Press, 1984.

[18] THE ATM FORUM. *ATM User-Network Interface (UNI) Specification Version 3.1.* Prentice Hall PTR, Upper Saddle River, NJ, 1995.

[19] M. H. WILLEBEEK-LEMAIR and A. P. REEVES. Strategies for Dynamic Load Balancing on Highly Parallel Computers. *IEEE Trans. on Parallel and Distributed Systems*, 4(9):979–993, 1993.

[20] C.-Z. XU and F. C. LAU. Iterative Dynamic Load Balancing in Multicomputers. *J. Opl. Res. Soc.*, 45(7):786–796, 1994.

[21] C.-Z. XU and F. C. LAU. Optimal Parameters for Load Balancing with the Diffusion Method in Mesh Networks. *Parallel Processing Letters*, 4(1–2):139–148, 1994.

2.3 Classification of Load Models

by Christian Röder

2.3.1 Introduction

Load distribution algorithms base their decisions on information about applications provided before execution time and/or on online observation of system resources. More sophisticated load distribution techniques take results of previous decisions into account when deriving new decisions[1]. During online observation load information is measured and collected in the system or the application. The central questions concerning these observations are: What components of the execution environment and the applications should a load distribution facility observe? When should the measurements be initiated? How can the results of the measurements be evaluated and linked together with application requirements to conclude reasonable decisions? Answering these questions is an integral part of the load distribution design, since they focus on the representation, computation and evaluation of load. The construction of a viable *load model* which is used to determine the load is dictated by several constraints: (1) the overall design of the load distribution facility, i.e., whether it is application or system integrated; (2) the system model which is used when designing the load distribution facility (e.g., queueing network models, graph models); (3) the entities which should be mapped and/or redistributed by the load distribution facility (e.g., processes as non-partitionable entities, single procedures of processes, threads, database requests, data objects to be processed and/or communicated) and (4) the expected performance improvements as seen from the users' or system's point of view.

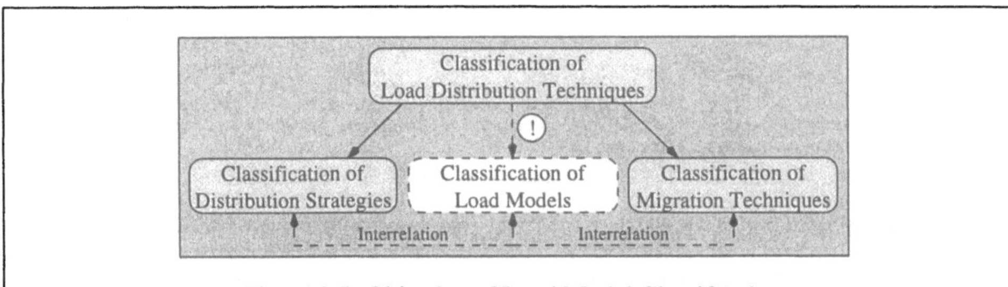

Figure 2.5: Objective of Load Model Classification.

One may find a variety of classification frameworks about load distribution algorithms and their policies [5; 14]. But generally, the computation and representation of load is limited to the selection of popular load index examples. Resource loads are among the most rapidly changing parts of the system state. The efficiency of load distribution techniques strongly depends on the values which are used to represent the load. Thus, from the load manager's point of view a well known general problem is to find a reasonable tradeoff between the expressiveness of the load model, i.e., the accuracy and amount of information used to determine the load, and the overhead introduced with measurement, computation and propagation of load values. In contrast to previous work we will not only list popular examples. We rather filter general characteristics of practical

---

[1] adaptive load distribution

load models for both system and application integrated load distribution techniques (Figure 2.5). The main issue of this section is to propose a classification framework for load models. The classification is useful for different purposes. It might be used for identifying already existing load models. On the other hand, it could serve as a guideline for a separate design of new load models in the context of load distribution algorithms.

The remainder of this section is organized as follows. 2.3.2 gives insight into widely accepted load models and their measurable attributes. By combining the measured values and the application requirements in a predefined manner the resulting value can be used by the load manager to determine the most suitable targets to execute the application entities. In 2.3.3 we introduce a classification of load models. The classification groups commonly accepted load indices into classes and aims at comparing already existing approaches and designing new techniques. In 2.3.4 we summarize the results and conclude.

2.3.2 Load Models and Load Indices

In concrete scenarios load is represented by load models and load values are calculated according to load indices. 2.3.2.1 introduces the definitions of load indices and load models and discusses the two terms. In 2.3.2.2 desirable properties when specifying a load index are reviewed and criteria are proposed which should guide the retrieving of load information. In 2.3.2.3 we list further information used for computing the load and provide a list of widely accepted formulas for the computation. Finally, in 2.3.2.4 we list popular examples of widely accepted load models.

2.3.2.1 Definition

To quantify the concept of load, FERRARI and ZHOU [9] use a *load index*, which they define as *a non-negative variable taking on a zero value if a (system) resource is idle, and increasing positive values as the load at the resource increases.* A load index is a quantitative measure of the load of targets. In their context of system integrated load distribution this definition focuses on and is limited to system resources. Within application integrated load distribution techniques this terminology is often adopted, but information about the utilization of system resources is often ignored. In this case, the load index is calculated taking only application characteristics into account (e.g., data objects to be processed or communicated, number of loop iterations). Generally, two different types of measurable values exist which indicate the *actual* or *average* utilization of a resource. We subdivide the type of load indices according to the time dependency of the measured values into two major groups: (1) An *instantaneous load index* is a load index which expresses the load value at dedicated snapshots of time. (2) A *comprehensive load index* is a load index which represents a certain number of load determining activities within a dedicated time period. The length of the time period is essential for computing the load value.

LUDWIG [14] defines the term *load model* as *the set of all measured values which are used for load evaluation, together with their weights and the kind of their combination.* Our classification in 2.3.3 will focus on those three items: the values used, i.e., the particular load indices, their weights and the type of their combinations. The resulting value of the combination can be viewed as a *global load index* (2.3.2.3). This global load index is propagated to other components of the load manager and is used for further load evaluation (i.e., the comparison of load situations of different targets). Obviously, several viable possibilities exist to define load models covering main sources of contention of the application and/or the system. Analogous to the distinction between system and application integrated load distribution we may refine the above definition by distinguishing between *system* and *application related load models*. Additionally, it is also possible to base load distribution decisions on load values measured in or provided by both the system and the application. We will call this a *hybrid load model*.

Note that there exist a large number of parallel applications which result in perfect load balance without explicitly measuring load indices. The load distribution is inherently formulated in the application. The used load index is of boolean type and can be viewed as the most primitive application related load model. The two values indicate whether there *is work* or there *is no work* actually performed. For example, to avoid an uneven assignment of units of work to targets, many shared memory programming systems use a central work queue from which idle targets remove units of work. A central work queue facilitates a dynamic, even distribution of load among the targets and ensures that no target is idle while there is still work to be done [19; 16].

2.3.2.2 Selection of Load Indices

As already mentioned in 2.3.1, measurements show that the performance benefits of load distribution are strongly dependent upon the selection of representative load indices [10; 12; 3]. FERRARI and ZHOU [9] specified a number of desirable properties in order to evaluate the quality of a load index. These criteria, in turn, depend on the performance index that is to be optimized by balancing the loads. From the system's point of view an increase of system throughput might be the primary goal whereas from the users' point of view a decrease of application runtime is of primary interest. Although the criteria were listed especially for system integrated load distribution they are easily extended to be usable for application related load models. A good load index should:

1. be able to reflect the qualitative estimates of the current load imposed on the target (expressions like "heavily loaded" become meaningful),
2. be usable to predict the load in the near future, since the performance index will be affected more by the future load than by the present load (it is often assumed that a currently heavily loaded resource will also be heavily loaded in the near future),
3. be relatively stable, i.e., high-frequency fluctuations in the load should be discounted or even ignored, thus reducing fluctuations in the control loop of the load manager,
4. have a simple (ideally, linear) relationship with the performance index, so that its value can be easily translated into that of the expected performance index.

We note that one general characteristic is often implicitly assumed in both application and system integrated load distribution: since load is produced by some entity the load model reflects its relation to this entity (or the resource requirements of the entities). For example, it is often assumed that execution time of processes relies on a single source of contention. If the CPU is identified as the main source of contention then CPU related values should be captured by the load model. If the same load model is used for I/O bound applications the load indices will not represent the load on the responsible source of contention. From our point of view we additionally introduce the following criterion to evaluate the quality of the load model:

5. The load model should express its relation to the load producing entities.

The above topics are only concerned with the desirable qualities of load indices. They will only indirectly influence our classification since those are mainly characteristics of the load evaluation than of the load model itself.

The influence of determining the values by measurement and sampling is left unattended. But the overhead introduced by measurement is also critical to the efficiency of the load manager and to the influence on the system. If the amount of measured load indices and the sampling rate increases the influence on the whole system increases as well. More sophisticated descriptions are typically more difficult to measure and to broadcast, thereby increasing the load distribution overhead. For example, EAGER *et al.* [7] noticed, when using different kinds of scheduling algorithms, that extremely simple load sharing policies using a small amount of information perform nearly as well as more complex policies that utilize more information. There has to be a viable tradeoff between the accuracy of information and the overhead of measurement. MEHRA and WAH [17] introduced some more criteria especially concerning this topic. From their point of view a load index should satisfy the following criteria:

6. low overhead of measurement, which implies that measurements can be performed frequently, yielding up-to-date information about load,
7. ability to represent the loads on all resources of contention, e.g., load information about CPU are only sufficient for compute bound tasks,
8. measurable and controllable independently of each other, i.e., mutually independent load indices are necessary.

So far we have described the desirable properties of load indices and the specified criteria for the measurement of load information. Finally, we may state when the measurements and the propagation of load information should be initiated. It might be questionable to include this topic into the classification of load models since this is a policy which is defined by the load distribution algorithm. But we think that it is not only important to define the constituting attributes of the load model but also its interrelation with other components of the load manager. The *initiation of measurement* could either be *on demand*, i.e., as a result from an external request, *periodically* after a *fixed time quantum* or *adaptive* to changing load situations. The former is used when the decision unit recognizes a major change in the load situation of one target and needs to find another target to balance the load. This approach is sufficient if it can be assumed that major changes in the load situation do not occur frequently. The second approach is the general case. It is used when a load manager cycles through its control loop in discrete time steps. Variable time steps are used to reduce the influence on an already overloaded target. The frequency of measurement is adapted to the load situations. Remember that some parallel applications achieve almost perfect load balance without explicitly using a load model. They use this kind of adaptivity since the time at which pieces of work are scheduled to each task depends on the load situation.

Similarly, we might state that the *propagation* of the global load information to other load distribution components may follow different strategies. It could either be done *on demand, directly after* the measurement and the computation of the global load index or *adaptive*, i.e., only if major changes in the load situation occur. For example, the latter technique is used to reduce the influence on the communication network significantly.

2.3.2.3 Computing the "System Load"

Remember that according to Ludwig's definition of load models (2.3.2.1) not only the measured load indices are important for load evaluation, but also their *weights* and the kind of their *combination*. In this subsection we will review some common techniques for determining the weight factors and combining the values measured at different sources of contention.

We will start with the determination of the weight factors. The main objective of the weight factors is to scale the measured values to a common base and to describe the relative impact of a single load value when computing the global load index. For example, the first objective covers situations in which the number of tasks in the run queue is measured (e.g., number of tasks is equal to 5) and combined with the number of free memory pages (e.g., the number of free pages is about 100). The relative impact of each value depends on the resource demands of the applications. Determining the relative impact of load values on the global load is the most critical factor when determining the weight factors. Several techniques and experiments have been performed to answer this question. The most popular approach is to predict the resource usage of applications [6; 13; 17]. For example, resource usage patterns might be predicted before execution of the applications as a byproduct of compile time analysis. Dynamic information like the input data size which are neglected in this approach might be collected and recorded during several previous runs of the applications. Additionally, EVANS and KESSLER [8] state that if the load values could be modeled in a time sensitive manner, then the instantaneous loadings of the resources could be estimated. Thus, not only the load values vary over time but also their weights may vary.

In recent years the utilization of distributed systems consisting of coupled heterogeneous workstations became more and more important as a research platform for parallel computing. In contrast to homogeneous multiprocessors and multicomputers two main characteristics influence the computation of load: heterogeneity and the usage as a time sharing system. To compute the load of one site the influence of external user processes has to be considered [2; 20; 22; 18]. External user loads in time sharing systems may have significant impact on the performance of parallel applications. Considering performance heterogeneity of the processing elements is essential if the load values of the sites have to be compared [4; 11]. The most common approach is to define a measure for the base performance of one dedicated site and to scale the performance of the other sites according to this *architecture factor*. But, heterogeneity might for example also be expressed by the presence or absence of local disks. It might be useful that for I/O bound applications the presence of local disks is more important than the performance of the processor. In general, static information about the hardware architecture of different sites is used as part of the weight factors.

Finally, we want to review two common techniques used for the combination of load values and adopt the notions of KUNZ [12]. n load indices $d_1, d_2, \cdots, d_n$ can be combined in a number of ways to determine the status of a processing element. The following examples give some general hints:

- Definition of a function of the form: $f(d_1, d_2, \cdots, d_n) := a_1 * d_1 + a_2 * d_2 + \cdots + a_n * d_n$, where the coefficients a_i are the weight factors.
- Combination of different load indices by the boolean operations AND, OR and NOT. Using AND indicates that a target is considered overloaded when all descriptors have values above their respective threshold. Using OR indicates that a target is considered overloaded whenever at least one of the descriptors has a value above its respective threshold.

We state that both the determination of the weight factors and the combination of the values are of importance when computing the load of targets. We further state that very simple load models seem to be sufficient for homogeneous systems executing compute bound applications. In case of heterogeneous systems and if other classes of applications (I/O bound, communication bound) have to be considered, the usage of architectural factors seems to be the most natural approach. But it is still not clear whether more complex load models (e.g., load models considering the time variance of resource requirements) would achieve superior distribution decisions in this situation. Therefore, we will include the influence of static information into the classification of the load models.

2.3.2.4 Examples

A wide variety of load indices has been used in the literature covering different aspects of the applications, the runtime and the hardware system. Remember that we distinguish between application and system related load models. We first list and discuss some examples for load indices used in system related load models and will continue with load indices used in application related load models. Hybrid load models use load indices located in both load models.

In system related load models sources of load information are located in the operating system. Both data structures (e.g., resource queues) and algorithms (e.g., local scheduling) are used for load value measurements. Load information belongs to all or only part of the main sources of contention: (1) CPU, (2) main memory, (3) secondary storage and (4) the interconnection network links. Many existing methods for computing load indices use simplified queuing models of computer systems. Almost all implemented systems use a function known as the UNIX-style load average (hereafter, load average), which is an exponentially smoothed average of the total number of processes competing for CPU [17]. Load average meaningfully compares loading situations across configurationally identical sites, but fails when the distributed system is heterogeneous with respect to performance[2]. A fundamental problem with the traditional load average function is that it completely ignores resources other than the CPU. Therefore, while it may be reasonable to predict performance of purely compute bound tasks, its utility is questionable for tasks that use other resources of contention: memory, disk and network (see discussion in 2.3.2.3). Although measuring the utilization levels of resources other than the CPU may not require any hardware modifications, several of these values are unsuitable for inclusion in a load model because the overhead of estimating their values precludes frequent sampling. Even if this can be eliminated, we are still left with a fairly large set of mutually dependent variables; for example disk traffic is affected by the number of page swaps and process swaps. Other values, such as rate of data transfer are fixed quantities for a given configuration and affect only the (fixed) coefficient of the load indices.

Typical application information used in load distribution reflects the computational and communication costs of each task. The costs in turn depend on several aspects, e.g., the number of loop iterations in a given algorithm or the distribution and the size of the processed data space.

LUDWIG [14] developed a system integrated load distribution facility for a homogeneous multiprocessor system and proposes a list of possible load information values which covers both as-

---

[2]in contrast to heterogeneity with respect to code compatibility

pects, the system utilization and the resource demands of application tasks. Table 2.4 lists the values which are used by the decision strategy to evaluate the load situation on each target.

node values	process values
processor idle time	time using CPU
process ready queue	time in process ready queue
sending queue	time in sending queue (to node x)
receiving queue	time in receiving queue (from node x)
free memory	used memory
amount of data sent	amount of data sent
amount of data received	amount of data received
number of messages sent	number of messages sent
number of messages received	number of messages received
	number of page faults

Table 2.4: Examples of load indices according to LUDWIG [14]

One obvious restriction implied in Table 2.4 has to be mentioned: the I/O load on each processing element is left unattended. In the context of Ludwig's experiments I/O loads could be neglected since only compute, memory and communication intensive applications had been investigated. Additionally, in SPMD style applications on homogeneous multiprocessors all tasks generally access an equal sized portion of the data on secondary storage. But if, for example, the environment consists of coupled heterogeneous workstations, in which some of the workstations might be diskless, the load on the processing elements imposed by I/O operations might vary significantly. In this case, the I/O load might strongly influence the runtime of I/O intensive applications and the focus has to be on the I/O behavior of the application tasks [10; 15]. Examples for values representing the I/O are the number of disk accesses or the size of the accessed data objects. A possible weight factor influencing the I/O load is the data distribution, i.e., the distance between the data and the processes accessing the data.

Finally, the interaction with the "non-hardware related" software components (e.g., specialized servers) or algorithms of the operating system might influence the behavior and runtime of the applications. Especially, the efficiency of the local CPU scheduling, local memory management algorithms, local I/O management and the software components responsible for intercommunication (locally or remote) might determine the runtime of executing tasks. KUNZ [12] proposed some load indices related to software components and operating system. The rate of CPU context switches and rate of system calls are examples for operating system related load indices. Since those values are determined within the runtime system of the applications we will group them into the system related load model.

2.3.3 Classification of Load Models

In 2.3.2.4 we listed commonly used examples of load indices and their embedding into load models. In contrast to the investigation of their influence on the efficiency of the overall load distribution facility we will now propose a new classification of load models. Generally speaking, the goal of a classification framework is to group heterogeneous objects into classes according to meaningful object characteristics. In our case, the objects of concern are load indices and the load models derived from them. In 2.3.3.1 we will review two approaches for the classification. We will discuss the approaches and show that both are limited in their extent. We will summarize 2.3.2 to derive the classification framework in 2.3.3.2. Finally, we will apply the classification to a system integrated scheduler for parallel jobs in heterogeneous workstation clusters.

2.3.3.1 Reviewing Approaches for Load Model Classifications

LUDWIG [14] states that a load model is composed of two submodels: the model used to determine the load on targets and the model used to determine the load of entities. In analogy, we distinguish between system and application related load models. Table 2.4 lists examples of processing elements and task related load informations. Ludwig investigated four submodels to determine the load on the processing elements and one fixed submodel for task loads. For his experiments he used values indicating (1) the idle time of the processors, (2) the length of the process ready queue, (3) an equal weighted combination of both values and (4) additionally the length of the receiving message queue. To determine the load of the tasks he combined the single weighted compute time and the double weighted waiting time of each task. Measurements are initiated at fixed time steps and the propagation is performed immediately after the measurements.

As already mentioned in 2.3.2.4 Ludwig proposed his classification for a system integrated load manager. The scope is limited to a dedicated hardware environment (homogeneous multiprocessor system) and usage model (exclusive access to space-shared partitions). In this sense, the classification of Ludwig is only a special instance of our more general classification proposed in 2.3.3.2.

A classification of *stochastic* load models was proposed by SCHNEKENBURGER [18]. The issue of stochastic load models and the classification aims at describing the external loads in a network of coupled workstations. The term *external load* is used to describe the load – additionally to the application load – which has to be performed by the system and was imposed on the system by other users. Four main classification criteria have been proposed:

static versus dynamic load models: static models assume no change in external loads, dynamic models are characterized by the arrival process of user submitted jobs and their workload distribution,

homogeneous versus heterogeneous load models: homogeneous load models are used if the system is assumed to be homogeneous and the resources are assumed to be identically distributed, else a heterogeneous load model is viable,

independent versus dependent load models (mainly used for system integrated load distribution): dependent load models take decisions of the load distribution into account, i.e., the increase of the load on one target causes a decrease of the load of another target if entities are migrated between these two targets,

quantitative versus qualitative load models: quantitatively, the application's resource requirements can fully be described by the work to be performed by the resource; qualitative load models use further attributes like priorities to evaluate the global load index.

In contrast to Ludwig's classification the load model characterization of Schnekenburger is of more general nature. But the classification lacks some general characteristics. For example, the types of measured values (instantaneous versus comprehensive) are left unattended.

2.3.3.2   Load Model Classification

Most system integrated load distribution techniques lack detailed knowledge about the occurrence of resource demands during the runtime of applications. In this case, load distribution is often limited to the initial mapping of applications. The mapping may be guided by information gathered during compile time, by knowledge about the runtime of applications recorded during previous runs or by specifications about precedence and communication constraints given by the application developers. This information is no load index in the above sense since there is no real measurement. On the other side, since it is used to reach distribution decisions we might state that this type of pre-runtime information is called a *static load model*. In contrast to that we will call the load model *dynamic* if online load information is measured.

To summarize the general characteristics proposed in 2.3.2 we propose a new classification framework for dynamic load models. The guidelines for the classification development are highlighted in Figure 2.6. The quality criteria listed in 2.3.2.2 are not part of the classification since they are no characteristics of the load model itself. Table 2.5 lists the characteristics and the categories for all relevant aspect of the classification. The desirable properties to evaluate the quality of the load model only indirectly influence this framework. Each table entry will be explained and discussed in the following subsections. For each topic at least one example will be provided.

Load Index Properties: We subdivide the table entry for the load model entities into two further categories: the general characteristics of the measured values, i.e., the general load index properties, and optional characteristics when combining multi-dimensional load indices, i.e., the load value composition. We sill start with the properties of the load indices:

index dimension: If only a single value is used to determine the load we call the load index *single-dimensional* (e.g., the average CPU utilization is the only value representing the load). If at least two values are combined to compute the load we call the load model *multi-dimensional* (e.g., average CPU utilization + average memory utilization).

index type: The type of load indices describes the main sources which are utilized to determine the load. Analogous to the distinction between CPU-, memory-, I/O- and communication-bound applications the type is described in terms of *CPU, MEM, I/O* and *NET*. For example, a system integrated load manager measures the length of the process ready queue or an application integrated load manager measures the number of loop iterations within a time interval. In both cases the values represent computation related load indices. Thus, we state

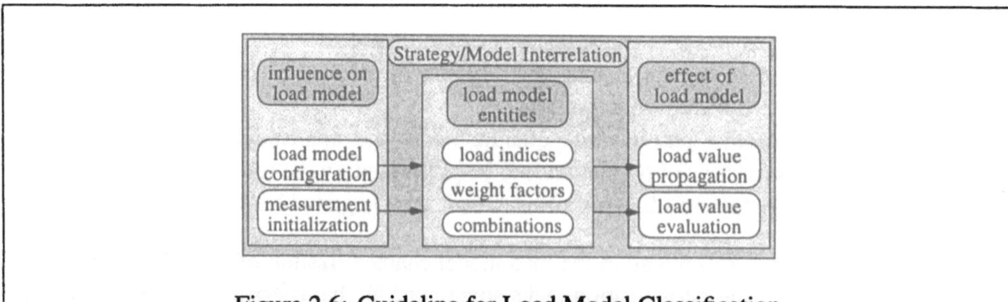

Figure 2.6: Guideline for Load Model Classification.

		Characteristic	Categories
Load Model Entities		**Load Index Properties**	
		index dimension:	- single-dimensional - multi-dimensional
		index type:	- CPU (computation) - MEM (main memory) - I/O (secondary storage) - NET (communication) - service - application defined
		aggregation time:	- instantaneous - comprehensive - component dependent
		aggregation space:	- system related - application related - hybrid (application and system)
		Load Value Composition	
		index combination:	- weighted linear combination - boolean selection - complex formula - none
		weight selection:	- pre-runtime application information - previous decision results - system characteristics - not applicable
Model Usage Policies		**index measurement:**	- on demand - periodically fixed - implicit
		index propagation:	- on demand - after measurement - adaptive
		model adaptivity:	- fixed - adaptive weighting - adaptive typing - adaptive combination

Table 2.5: Classification of load models

that the index type is CPU. The type of the load indices is *service* if applications request service from an outside component (e.g., client-server applications, data base systems, operating systems). A typical example is the number of system calls of the application. Finally, some applications define their own index types. For example, image processing applications may use the number of pixels processed per time to indicate the load situation.

aggregation time: Load indices which represent the load at a dedicated timestep are called *instantaneous* (e.g., number of free memory pages at time t_i). Load indices which represent the load for a certain period of time are called *comprehensive* load indices (e.g., average CPU utilization during time period Δt). The measured values in multi-dimensional load indices might be of both types. We will abbreviate this by *component dependent* instead of requiring that the two characteristics have to be assigned to each single value.

aggregation space: The load model is called *system related* if the measured load values are located within the hardware or operating system (e.g., the number of tasks in the ready queue of a processor). The load model is called *application related* if measurements only cover application relevant load values (e.g., number of loop iterations within a given time period). If information is utilized from both sources, the application and the system we call the load model *hybrid*.

Load Value Composition: The second part of the load model classification covers the computation of the global load index which will be propagated to the other load distribution components.

index combination: Several approaches exist for the combination of the vector elements in a multi-dimensional load index. The most common approach is the *weighted linear combination* of the vector elements. Another alternative is *boolean combination*. The vector elements are combined by the boolean operators AND, OR and NOT. A more *complex formula* might also be used if, for example, the vector elements are not mutually independent.

weight selection: The relative impact of different measured load values is determined by their respective weight factors. Weight factors might be selected according to the following approaches: (1) pre-runtime application information about its resource requirements is known in advance, (2) *previous decision results* of the load distribution strategy or (3) architectural *system characteristics* determine the weight factors.

Model Usage Policies: The interrelation between the load model and the load distribution strategies determines the usage and effectiveness of the load model. Both the model and the strategies are mutually dependent.

index measurement: The initiation of load value measurements is called *on demand* if measurement is initiated by a request from a load distribution component (e.g., polling of targets if entities have to be migrated). The initiation is called *periodically fixed* if load values are sampled after fixed time quantum without influencing the sampling period (e.g., constantly measuring the number of I/O operations and average number of disk accesses after 1 sec). The initiation is called *implicit* if the sampling period changes according to changing load situations (e.g., when using a central work queue).

index propagation: The propagation of load values could either be *on demand* (e.g., when searching for a suitable target to which load can be migrated), *after measurement* (e.g., broadcasting of values whenever new values are measured) or *adaptive*, i.e., only major changes of the load situations are propagated to other load distribution components. Adaptive index propagation assumes some kind of pre-evaluation of load situations.

model adaptivity: A *fixed* load model is used if all three components, i.e., the definition of the measured values, the weights and the type of combination, are fixed during the execution time of the load manager. For example, *fixed* weighting is used if the relative impact of the vector elements is assumed to be fixed during the whole execution time (e.g., double weighted CPU utilization, single weighted number of free memory pages). We call the load model *adaptive* if one of these components is dynamically reconfigurable during the execution time (e.g., skipping from I/O bound to CPU bound execution phases of applications causes an online reconfiguration of the weight factors). To be more precise, we introduce the terms *adaptive weighting*, *adaptive typing* and *adaptive combination* to indicate whether the weight factors, the measured index types or the combinations might change. For example, *adaptive* weighting is used if the relative impact of the measured values changes during the execution time (e.g., consider changing resource demands during dedicated periods of time).

2.3.3.3 Applying the Classification

In this section we review a load distribution technique for scheduling parallel jobs on a cluster of heterogeneous[3] workstations. Although the main objective of this book is to investigate application orientied load distribution techniques, this example demonstrates that the classification is also applicable for system integrated load distribution techniques.

Load distribution algorithms for heterogeneous environments have to consider the different processing speeds of the workstations, i.e., the targets. WEINRIB and SHENKER [21] proposed a commonly used load function to manage the load distribution. The tasks of a parallel application, i.e., the entities, are allocated to that target i with the minimum value of the load function, which is given by:

$$\varphi_i = \frac{n_i}{\mu_i \times f_i} + \frac{1}{\mu_i}$$

where n_i is the number of entities at the ready queue of target i, f_i is the fraction of the target i idle time during the last minute and μ_i is the target i speed (e.g. MIPS, MFLOPS, etc.). A load distribution algorithm will select the target i with the minimum value φ_i when a new entity has to be assigned. The algorithm will select the fastest target if several targets could be selected. If all targets are busy, the algorithm selects that target which offers the best combination of minimum load (represented by n_i and f_i) and largest speed (μ_i).

ALMEIDA *et al.* [1] use this load function when designing a distributed scheduler for parallel jobs. The global load index is periodically determined at fixed pre-defined timesteps. To reduce the number of message transmissions the global load index (i.e., the value of the load function) is propagated to the distribution strategy only if the load varies significantly. Table 2.6 summarizes the used load model. (Table 2.6 is a comprehensive version of the classification scheme proposed in Table 2.5 on page 43. This style will be used throughout the rest of the document).

2.3.4 Summary

In 2.3.2.2 we listed important quality criteria when designing a load model for load distribution techniques. Furthermore, we listed and discussed a wide variety of popular load indices both for

---

[3]heterogeneity with respect to different performance of the targets

Characteristic	Categories
Load Index Properties	
index dimension:	multi-dimensional (dim = 2)
index type:	CPU (computation)
aggregation time:	instantaneous, comprehensive
aggregation space:	system related
Load Value Composition	
index combination:	complex formula
weight selection:	system characteristics
Model Usage Policies	
index measurement:	periodically fixed
index propagation:	adaptive
model adaptivity:	static

Table 2.6: Load Model for Scheduling Parallel Jobs in Heterogeneous Environments

system as well as for application integrated load distribution. The focus of the discussion was to refine the notions of the terms load index and load model. In 2.3.3 we identified and filtered general characteristics of commonly used load models and proposed a new classification framework of those models. This classification should guide the developer of future load distribution techniques to classify their load models.

References

[1] V. ALMEIDA, J. ARABE, E. LOURES and G. RIMOLO. Scheduling Parallel Jobs on a Cluster of Heterogeneous Workstations. In *High Performance Computing Conference '94*, pages 103–108, Sep 1994.

[2] W. BECKER. Dynamic Balancing Complex Workload in Workstation Networks – Challange, Concepts and Experience. In B. HERTZBERGER and G. SERAZZI, editors, *HPCN – Europe*, volume LNCS 919, pages 407–412, May 3-5 1995.

[3] L. CABRERA. The Influence of Workload on Load Balancing Strategies. In *1986 Summer USENIX Conference*, pages 446–458, June 1986.

[4] J. CASAS, R. KONURU, S. OTTO, R. PROUTY and J. WALPOLE. Adaptive Load Migration Systems for PVM. In *Proc. Supercomputing'94*, pages 390–399, Nov. 14-18 1994.

[5] T. L. CASAVANT and J. G. KUHL. A Taxonomy of Scheduling in General-Purpose Distributed Computing Systems. In *IEEE Trans. Softw. Eng.*, pages 141–154, Feb 1988.

[6] M. V. DEVARAKONDA and R. K. IYER. Predictability of Process Resource Usage: A Measurement Study on UNIX. In *IEEE Trans. Softw. Eng.*, volume 15(12), pages 1579–1586, dec 1989.

[7] D. L. EAGER, E. D. LAZOWSKA and J. ZAHORJAN. Adaptive Load Sharing in Homogeneous Distributed Systems. In *IEEE Trans. Softw. Eng.*, volume SE-12, pages 662–675, May 1986.

[8] J. D. EVANS and R. R. KESSLER. Allocation of Parallel Programs With Time Variant Resource Requirements. In *1993 Int. Conf. on Parallel Processing, Vol. III*, pages 271–275. CRC Press, Inc., 1993.

[9] D. FERRARI and S. ZHOU. An Empirical Investigation of Load Indices for Load Balancing Applications. In *Performance'87, 12th Int. Symp. on Computer Performance Modeling, Measurement and Evaluation*, pages 515–528, 1987.

[10] A. HAĆ and T. J. JOHNSON. Sensitivity Study of the Load Balancing Algorithm in a Distributed System. *Journal of Parallel and Distributed Computing*, 10:85–89, 1990.

[11] M. HAMDI and C. LEE. Dynamic Load Balancing of Data Parallel Applications in a Distributed Network. In *ACM Int. Conf. on Supercomputer*, pages 170–179, 1995.

[12] T. KUNZ. The Influence of Different Workload Descriptions on a Heuristic Load Balancing Sheme. In *IEEE Trans. Softw. Eng.*, volume 17(7), pages 725–730, July 1991.

[13] W. LELAND and T. OTT. Load Balancing Heuristics and Process Behaviour. In *ACM SIGMETRICS Conf. on Measurement and Modelling of CS*, pages 54–69, May 1986.

[14] T. LUDWIG. *Automatische Lastverwaltung für Parallelrechner.* Reihe Informatik, Band 94. BI-Wissenschaftsverlag, 1993.

[15] S. MAJUMDAR and Y. M. LEUNG. Characterization of Applications with I/O for Processor Scheduling in Multi-programmed Parallel Systems. In *6th IEEE Symp. on Parallel and Distributed Processing*, pages 298–307, Oct. 26-29 1994.

[16] E. MARKATOS and T. J. LEBLANC. Using Processor Affinity in Loop Scheduling on Shared Memory Multiprocessors. In *IEEE Trans. on Parallel and Distributed Systems*, volume 5(4), April 1994.

[17] P. MEHRA and B. WAH. Automated Learning of Workload Measures for Load Balancing on a Distributed System. In *1993 Int. Conf. on Parallel Processing, Vol. III*, pages 263–270. CRC Press, Inc., 1993.

[18] T. SCHNEKENBURGER. *Adaptive Lastverteilung für parallele Programme.* Dissertation, Techn. Univ. München, 1994.

[19] S. SUBRAMANIAM and D. L. EAGER. Affinity Scheduling of Unbalanced Workloads. In *Proc. Supercomputing'94*, pages 214–226, Nov. 14-18 1994.

[20] S. TURNER, L. NI and B. C. CHENG. Time and/or Space Sharing in a Workstation Cluster Environment. In *Proc. Supercomputing'94*, pages 630–639, Nov. 14-18 1994.

[21] A. WEINRIB and S. SHENKER. Greed is not enough: Adaptive Load Sharing in Large Heterogeneous Systems. In *Proc. of the IEEE INFOCOM*, 1988.

[22] Y. ZHANG, K. HAKOZAKI, H. KAMEDA and K. SHIMIZU. A Performance Comparision of Adaptive and Static Load Balancing in Heterogenous Distributed Systems. In *Proc. 28th Simulation Symposium*, pages 332–340. IEEE Comp. Soc. Press, 1995.

2.4 Migration Mechanisms

by Georg Stellner

A means to migrate parts of a computation between targets is required to balance the load of applications at runtime. The following section presents a taxonomy that allows to classify different approaches of migration mechanisms. This terminology is used in the last section to discuss examples of different migration mechanisms.

2.4.1 A Taxonomy for Migration Mechanisms

The taxonomy presented here is part of a more general terminology that is used to describe check-pointing mechanisms [10]. The discussion looks at migration components as independent parts of load distribution systems which is accessed through an interface. This interface is used by the strategy component to submit request to perform migrations. For many existing systems such an interface is not obvious. In these cases the existence of an interface is assumed and the classification is given accordingly.

As discussed previously (cf. section 2.1.1) the main objective of a load distribution system is to assign *entities* to *targets*. Hence, the first distinction between migration mechanisms is the entity which is actually migrated between targets to level the load: the *migrant*. A great variety can be used as migrant in a migration mechanism. Two major classes of migrants can be distinguished, namely *active* and *passive* migrants. Whereas the former are active components of the system that have an execution thread the latter are only used by active components. Examples of active migrants are processes or threads. Passive migrants are for example data or objects[4]. In order to classify also more peculiar migration mechanisms further terms can be used to specify the migrant, such as transaction or file.

For the remainder of the document the term *migration set* will refer to the set of migrants which can be migrated in a single step. This is important if several migrants ought to be migrated in a single request. If the number of migrations within the migration set is limited then the request to migrate all migrants must be carried out sequentially in chunks of the maximum number of migrants in the migration set. This imposes more overhead than migrating all migrants simultaneously and consequently decreases the efficiency of the migration mechanism. Migration mechanisms that have a limitation on the number of migrants are referred to have an *limited migration set size* and mechanisms that do not impose any restrictions on the number of migrants have an *unlimited migration set size* respectively.

A further property of migration mechanisms is the support for *heterogeneity*. If migrants can only be moved between identical machines no support for heterogeneity is provided. Particularly for migration mechanisms using active migrants migration between heterogeneous machines is difficult as different machines usually have different processors with different instruction sets. Also the runtime environment of an active migrant must be established on the destination machine. Systems that support heterogeneous process migration are described in [13] and [9]. But also for mechanisms using passive migrants support for heterogeneity is difficult to achieve as it might be necessary to convert data types or data structures.

The request to migrate a given migration set is an asynchronous event to the computation of the application. After the request has been intercepted the migration system can either start immediately to migrate or must delay the migration until the application has reached a certain state where migration is permitted. In the former case the migration mechanism must take precautions to maintain a consistent state of the application also after migrants have been migrated. In the latter case migration is only permitted when the application is in a state that is consistent[5]. Migration mechanisms that can react immediately to incoming migration requests, the *migration initiation* is *immediate*. Otherwise the migration initiation is referred to as *delayed*.

The transfer of the data, which represents a migrant and is used on the destination to reconstruct the migrant, can be improved with different techniques:

Pre transfer media: A *pre transfer media* can be used to store the data before it is actually transfered over the network. Later on, the data is transfered over the network concurrently to the computation of the application on the source node. This reduces the time during which a migrant blocks the computation on the source node achieve consistency.

Compression: *Compression* techniques can be applied to reduce the amount of data that has to be transferred to migrate a migrant. This is beneficial if the time which is required to compress and uncompress the data is less than the time that is required to transfer the additional data without compression. A quantitative discussion about compression and its benefits can be found in [7].

Transfer Policy: The transfer policy determines how the data representing a migrant is transfered to the destination node. The data can either be transfered *completely* or *partially*. In contrast to a complete transfer where all the data describing a migrant is transfered as a whole, partial

---

[4]Objects here refer to objects in the sense of object oriented programming.

[5]For a discussion about consistent states in distributed systems refer to [2].

transfer techniques split the data into junks. The transfer of the junks is depending on the partial transfer technique as described below:

Lazy Copying: The *lazy copying* technique can be seen as *remote* demanded paging [14]. An initial minimal information set is migrated to the destination. Further parts which are necessary later on are requested from source nodes[6] and transfered to the current node.

Enhanced Lazy Copying: The *enhanced lazy copying* approach is quite similar to lazy copying, except that always a specific node, a so called *home node*, of a migrant serves the requests for further parts of a migrant. Therefore the home node manages the data that is associated with a migrant and the current node of a migrant keeps book about modified junks. Upon migration time these modified junks are flushed back to the home node [3].

Pre Copying: In *pre copying* a migrant must not be frozen during migration. If it is an active migrant it can continue with its computations. If it is a passive migrant it can still be used by active entities in operations. The migration is iterative, where in the first step the complete migrant is transfered to the destination node disregarding ongoing operations. In the next step all junks of a migrant which have been modified in the meantime on the source node are again transfered. This procedure is repeated until the amount of junks that have to be transfered is below a certain threshold. Then the migrant is frozen and the remaining modified junks are transfered a last time to the destination node [12].

A further important property of migration mechanisms is the degree of dependency from other nodes when migrations are being performed. This characteristic has been intensively discussed in the literature under the term *residual dependencies* [1; 3; 5; 12]. DOUGLIS and OUSTERHOUT define a residual dependency as: "an ongoing need for a host to maintain data structures or provide functionality for a process even after the process migrates away from the host" [3, p. 771]. In the framework of this book the definition is generalized. Here, a residual dependency occurs if a migrant needs services or resources from a node other than the one, where the migrant currently resides. In this case, residual dependencies may also occur prior to migration when resources on other nodes need to be reserved for potential migrants. At a closer look three different dependencies can be distinguished: *source node*, *destination node* and *home node* dependencies. Source node dependencies and home node dependencies typically evolve when partial transfer policies are used. In these cases junks of the data describing a migrant may be still located on a source node and must be sent to the destination node on request. Destination node dependencies are mostly due to the ease of implementation and efficiency. Instances of potential migrants are kept on possible destinations and in case of an actual migration just the current state must be transfered. Hence, the time to create an instance of the migrant can be saved [6].

From the users' point of view *transparency* is an important property of migration mechanisms [5]. Two different aspects can be distinguished: *migration transparency* and *source code transparency*. A migration mechanism is migration transparent if the migration does not become obvious for the user. Another facet of transparency is *source code transparency*. A migration mechanism offers source code transparency if the source code of existing applications does not need to be modified to benefit from migration. For migration mechanisms that support an existing

---

[6]This includes nodes where migrant resided before the most recent migration

parallel programming environment such as PVM or MPI are source code transparent if the sources of existing programs can be used without modifications. If however, the migration mechanism provides its own parallel programming environment, all applications must first be ported to that particular parallel programming environment and the migration mechanism is hence not source code transparent.

Migration Mechanisms	
Migrant	process/thread/data/object/transaction/file/...
Migration Set Size	limited/unlimited
Heterogeneity	yes/no
Migration Initiation	immediate/delayed
Pre Transfer Media	cache/memory/disk/none
Compression	yes/no
Transfer Policy	complete/ partial (lazy copying, enhanced lazy copying, pre copying)
Residual Dependencies	source node/destination node/home node/none
Migration Transparency	yes/no
Source Code Transparency	yes/no

Table 2.7: Taxonomy of Migration Mechanisms

2.4.2 Examples of Migration Mechanisms

Using the taxonomy defined in the previous section the following two sections introduce two sample migration mechanisms. The first environment allows the migration of processes between heterogeneous machines. The second mechanism is part of a distributed operating system.

2.4.2.1 Distributed C

Distributed C [8] is a parallel programming environment which allows to write parallel programs with an interprocess communication similar to ADA [11]. This is an implicit rendez-vous style communication. In addition to that the standard UNIX interprocess communication is also supported. Recently, a component that provides process migration has been added [9]. The key feature of this extension is that the migration of the processes is also possible between heterogeneous nodes.

To allow the migration of processes between heterogeneous machines a special compiler technique is used. Before actually compiling the program, it is transformed into a MIG-rateable intermediate code. This is a representation of the program for a special virtual machine. The virtual machine comprises a virtual processor, a memory manager and an address table to map branch targets. The code which makes up the virtual machine is provided in libraries which are linked to the application program.

Defining such a virtual machine makes it easy to determine the state of a process. It is defined by the contents of the data structures of the virtual machine. The migration of one or more processes is now carried out as follows.

During compilation time the compiler has inserted has inserted special points into the code where migration is permitted. Whenever the computation comes across such a point it is tested whether or nor there is a request to migrate the process. If the process ought to be migrated then migration prolog is invoked. During this prolog, all file buffers are flushed and a skeleton process is started on the target. Then the state of the virtual machine is determined, transformed into a machine independent format and transfered to the skeleton process. The source process continues to execute as a shadow process to serve location depend-end system calls like file accesses. During a migration epilog on the target, the skeleton process receives the state information and initializes its data structures accordingly.

Migration Mechanisms	
Migrant	process
Migration Set Size	unlimited
Heterogeneity	yes
Migration Initiation	delayed
Pre Transfer Media	none
Compression	no
Transfer Policy	complete
Residual Dependencies	source node
Migration Transparency	yes
Source Code Transparency	yes

Table 2.8: Taxonomy of the Migration Mechanism of Distributed C

The classification of the migration mechanism of Distributed C is summarized in table 2.8.

2.4.2.2 Sprite

The Sprite project of the University of California at Berkeley implements a distributed operating system. The primary goal of the project is to hide the distribution of the machines and combine this with a UNIX compatible interface. Therefore a high degree of transparency has been provided. For this environment DOUGLIS has developed a process migration mechanism [4].

Processes in Sprite can be executed on any machine in the network. To provide the transparency Sprite distinguishes two different classes of system calls: location dependent and location independent calls. Whereas the result of the execution of the former depend on the node where they are executed the latter result of the latter is independent of the node. System calls of the first category must hence be executed on that node, where a process is started. Hence, Sprite executes these calls transparently in an RPC-like fashion on the machine where a process was started.

The migration in Sprite is integrated into the virtual memory management of the processes. Instead of using the local disk as the paging device, pages that need to be written to secondary storage are transfered to the node where a process was started and written to disk there. Similarly, if further pages are required on demand during run-time the request is forwarded to the node where the process was started and the required page is sent from there to the node that executes the process.

When a process ought to be migrated its current state can be easily determined. All pages that have been marked as "dirty" by the operating system on the node that executes a process need to be transfered to the node from where the process was started. The stack as a part of the address space is used to transfer the processor context. All registers are pushed onto the stack upon migration. Hence, these pages are also transfered to the node from where the process was started. Upon restart on the new node they are used to restore the processor context. According to the taxonomy this is an enhanced lazy copying transfer policy. Using the node from where a process was started to serve location dependent system calls and as a part of the local virtual memory manager causes residual dependencies from the home node. As the processor context is directly loaded into the processor on the destination node processes can only be migrated between homogeneous nodes.

Migration Mechanisms	
Migrant	process
Migration Set Size	unlimited
Heterogeneity	no
Migration Initiation	immediate
Pre Transfer Media	none
Compression	no
Transfer Policy	partial (enhanced lazy copying)
Residual Dependencies	home node
Migration Transparency	yes
Source Code Transparency	yes

Table 2.9: Taxonomy of the Migration Mechanism of Sprite

Table 2.9 summarizes the properties of the migration component of Sprite.

References

[1] Y. ARTSY and R. FINKEL. Designing a Process Migration Facility — The Charlotte Experience. *IEEE Computer*, 22(9):47–56, September 1989.

[2] K. M. CHANDY and L. LAMPORT. Distributed Snapshots: Determining Global States of Distributed Systems. *ACM Trans. on Computer Systems*, 3(1):63–75, February 1985.

[3] F. DOUGLIS and J. OUSTERHOUT. Transparent Process Migration: Design Alternatives and the Sprite Implementation. *Software — Practice and Experience*, 21(8):757–785, August 1991.

[4] F. DOUGLIS. *Transparent Process Migration in the SPRITE Operating System*. Ph.d. thesis, Computer Science Division (EECS), University of California at Berkeley, CA, 1990.

[5] M. R. ESKICIOĞLU. Design Issues of Process Migration Facilities in Distributed Systems. *IEEE Technical Commitee on Operating Systems Newsletter*, 4(2):3–13, Winter 1989.

[6] R. KONURU, S. OTTO, R. PROUTY and J. WALPOLE. A User-Level Process Package for PVM. In *SHPCC'94*, pages 48–55, Knoxville, TN, May 1994. IEEE Computer Society Press.

[7] J. S. PLANK and K. LI. ickp: A Consistent Checkpointer for Multicomputers. *IEEE Parallel & Distributed Technology*, pages 62–67, Summer 1994.

[8] C. PLEIER. The Distributed C Programming Environment. Technical Report TUM-I9324, Technische Universität München, 80290 München, September 1993.

[9] C. PLEIER. Ein Verfahren zur Prozeßmigration in heterogenen UNIX-Rechnernetzen. *PIK — Praxis der Informationsverarbeitung und Kommunikation*, 18(2):75–81, April 1995.

[10] G. STELLNER. *Methoden zur Sicherungspunkterzeugung in parallelen und verteilten Systemen.* LRR-TUM Research Report Series. Verlag Shaker, Aachen, October 1996. Reprint of the Phd-thesis which has been submitted to and accepted by the Fakultät für Informatik der Technischen Universität München in July 1996.

[11] M. J. STRATFORD-COLLINS. *ADA: A Programmers Cconversion Course.* Ellis Horwood Limited, Chichester, 1982.

[12] M. M. THEIMER, K. A. LANTZ and D. R. CHERITON. Preemptable Remote Execution Facilities for the V-System. In *10th Symposium on Operating System Principles*, volume 19 of *ACM Operating System Review*, pages 2–12, New York, NY, December 1985. ACM Press.

[13] M. M. THEIMER and B. HAYES. Heterogeneous Process Migration by Recompilation. In *11th Int. Conf. on Distributed Computing Systems*, pages 18–25. IEEE Computer Society Press, 1991.

[14] E. R. ZAYAS. Attacking the Process Migration Bottleneck. In *11th Symposium on Operating System Principles*, volume 21 of *ACM Operating System Review*, pages 13–24, New York, 1987. ACM Press.

2.5 Exemplary Load Distribution Concepts: A Classification

by Thomas Schnekenburger

This section classifies different approaches to load distribution according to the previously proposed classification schemes. In the first part, three methods for assigning processes to nodes are classified:

- Non-preemptive, process-based load distribution by EAGER, LAZOWSKA, and ZAHORJAN [5],
- a microeconomic concept for dynamic assignment of communicating processes presented by FERGUSON, YEMINI, and NIKOLAOU [6], and
- process based load distribution using a physical analogy by HEISS and SCHMITZ [10; 9; 8].

The second part shows four domain specific approaches to load distribution:

- Load distribution for parallel query execution by RAHM and MAREK [13],
- load distribution for a distributed file system by HUA, LEE, and YOUNG [11; 12],
- data placement and load balancing in disk-arrays by SCHEUERMANN, WEIKUM, and ZABBACK [14; 16], and
- load balancing for parallel image processing algorithms by HAMDI and LEE [7].

Run-time environments for load distribution are a compromise between system-integrated and application-integrated load distribution. The third part shows three examples:

- The PARFORM platform by CAP and STRUMPEN [4; 3; 15] for parallel computing in a distributed workstation environment,
- DIME by WILLIAMS [17; 18] for parallel computing based on irregular meshes, and
- the HiCon model by BECKER [2; 1] for data-intensive applications in workstation networks.

Several of the presented methods do not use migration or do not describe the underlying migration mechanism. In that case, the migration mechanism classification is omitted.

If the given system covers several different approaches that cannot be classified by a single attribute according to the classification scheme, we use the term *multiple* to indicate that there is more than one possible attribute.

2.5.1 System Based Load Distribution

"Adaptive Load Sharing in Homogeneous Distributed Systems" by EAGER, LAZOWSKA, **and** ZAHORJAN [5]

EAGER et al. describe a concept for assignment of processes (called "tasks") to nodes of a homogeneous distributed system with the goal of improving response time of the entire system. In contradiction to the previous classifications, the concepts are called "adaptive" just to indicate that the load distribution strategy reacts on the actual system state.

Regarding the general classification (Table 2.10 on page 57), this work is *not domain specific*. The intent is *system-oriented* load distribution by *assignment*. Monitoring, strategy and mechanisms are intended to be integrated into the *operating system*. Processes are independent and each process may be executed on any node, hence the entire system can be classified as *location independent*. Load distribution is realized by *placement* of *processes to nodes*, i.e. processes are subproblems and nodes are targets.

The paper describes three different strategies: "random", "threshold", and "shortest". The system model has a *combinatorial* flavor. The target topology consists of *homogeneous, full connected nodes*. The entities (here: tasks) are *non-interacting*. Transfer of tasks is *systemwide* and *non-preemptive*. Whereas the "random" strategy uses no information exchange, "threshold" and "shortest" use a *systemwide* information space. The information scope is *partial*, because the "probe limit" restricts the number of nodes that are probed (asked for their actual load). Task transfers are coordinated only between the sender and the receiver of a task. Therefore, the decision structure is *distributed, cooperative* and *partial*. The "random" strategy is *static*, "threshold" and "shortest" are *dynamic* strategies. Task transfers are initiated by the *sender* of tasks. The algorithm adaptivity is *fixed*. Cost sensitivity of "threshold" and "shortest" is *partial*, because communication overhead for task transfers is neglected. The "random" strategy has no cost sensitivity. Stability is guaranteed by a "transfer limit" that restricts the number of times that a task can be transferred.

The load of individual nodes is characterized solely by the queue-length, i.e., the number of tasks that are currently running on a node. Therefore, there is a *single-dimensional* load model regarding the *CPU*. Aggregation time is *instantaneous*, because only the actual queue length is considered. The queue-length is a *system-related* aggregation space. Index measurement can be classified as *implicit* since it is updated whenever a task is assigned or terminated. Index propagation is *on demand* ("probe"). The load model adaptivity is *fixed*.

A migration mechanism is not needed because task assignment is non-preemptive.

"Microeconomic Algorithms for Load Balancing in Distributed Computer Systems" by FERGUSON, YEMINI, **and** NIKOLAOU [6]

FERGUSON et al. present concepts for load distribution that are drawn from microeconomics. Prices for resources are set by competition among processes (called "jobs"). Resource allocation decisions are made through auctions held by the processors.

Generally, we can classify this work as *not domain specific* and *system-oriented* (Table 2.10 on page 57). Load distribution is realized by *assignment*. Monitoring, strategy, and mechanisms are assumed to be integrated into the operating system. The system consists of processes that are allowed to migrate to other nodes. Generally, the system structure is assumed to be *location dependent*. Load distribution is realized by preemptive assignment of processes to nodes, i.e. nodes represent load distribution *targets* and processes represent *workspaces*. Subproblems of the application can be classified as *computation* (cf. section 2.1.3).

FERGUSON et al. use a *microeconomic* system model consisting of *heterogeneous nodes* connected by a point-to-point network that can be defined by an arbitrary *graph* (Table 2.11 on page 57). The application consists of *non-interacting tasks*. Transfer of tasks can be *preemptive*, but it is only possible to transfer tasks to *neighbors*, i.e. along the edges of the given point-to-point network. The computation of prices and biddings for resources is performed by "auctions" and "advertisements". For a given node, all neighbors to that node participate in "auctions" and "advertisement" for its local resources. Therefore, the information space is *neighbor* and the information scope is *partial*. The decision structure is obviously *distributed*, *competitive*, and *partial*, since only a subset of all nodes competes via auctions for the available resources. Actual prices for resources can be regarded as an actual load evaluation of a resource. Consequently, the given strategy is *dynamic*. Initiation of task transfers is implied by the result of advertisement and auctions for a resource. The strategy may be rated as *receiver-initiated*, because advertisement and auctions are always initialized by the provider of a resource. The actual strategy parameters are not changed, hence the adaptivity is *fixed*. Cost sensitivity is *high*, because all resources are integrated into the microeconomic model. Tasks have to "pay" for their transfers. Tasks that perform too many useless transfers just run out of "money". Nevertheless, the nuber of transfers depends on a suited cost rating for transfers so that stability control is rated as *partial*.

Regarding the load model (Table 2.12 on page 58), resource "prices" represent load indices for resources. The paper uses *CPU* prices as well as network (*NET*) prices. Both prices are combined to decide about task allocation. Consequently there is a *multi-dimensional* load index. Load indices are *comprehensive* as resource prices develop over time. The load indices are clearly system related, Furthermore, application related load indices, as for example the normalized job execution time of jobs assigned to a certain node, are used. The aggregation space can be classified accordingly as *hybrid*. The composition of load values is represented by the combination of prices that are computed by adding prices for individual resources. Therefore, index combination corresponds to a *weighted linear combination*. The weights of this linear combination are *system characteristics* such as the "processor speed parameter" and *pre-runtime application information* such as the normalized CPU service time of a task. Index measurement is *implicit* by computing prices for resources. Prices are propagated only when they have changed, i.e. index propagation is *adaptive*. The model adaptivity can be classified as *fixed*, because the formulas for computing new prices and resource costs are not changed.

Although FERGUSON et al. use a preemptive strategy, they do not describe the underlying migration mechanism.

"Decentralized Dynamic Load Balancing: The Particles Approach" by HEISS and SCHMITZ [10]

HEISS et al. (see also [8; 9]) present a decentralized approach to load distribution where the load distribution entities (processes) are modeled by "particles". The idea is to use analogies between load distribution and the flow of particles in a physical system. The load distribution algorithm basically consists of a combination of four forces: A "load balancing force" tries to assign entities (called "tasks") to "the fastest" node, a "communication force" tries to reduce the distance between intensively communicating tasks, a "dispersion force" tries to locate tasks of the same parallel program to different nodes in order to increase parallelism, and the "damping force" tries to attract already assigned tasks in order to reduce the number of migrations.

Regarding the general classification (Table 2.10), the work of HEISS et al. is *not domain specific*. Although HEISS et al. emphasize the specific load distribution strategy and not the implementation of mechanisms, the work can be classified as *system-related*, because the distribution entities are processes (here called "tasks") of parallel programs. Load distribution is done by *assignment*. Monitoring, strategy and mechanisms are intended to be integrated into the *operating system*. Process migration is used, i.e. the problem structure is generally *location dependent*. Processes represent workspaces that are assigned to nodes. There is no description of the implementation of "tasks". Therefore, we classify subproblems just as *computation* that is performed by processes.

The strategy model (Table 2.11) is a *physical* model consisting of *heterogeneous* nodes that are connected by point-to-point interconnections forming an arbitrary *graph*. Applications are parallel programs with an arbitrary "task interaction graph" that describes communication channels and the communication intensity for each channel. Hence we can classify the entity topology as a *mesh*. Entities (here: tasks) are transferred only between *neighbored* nodes. Running tasks can be migrated, i.e. task assignment is *preemptive*. According to the physical model, information exchange is only with *neighbors* and information scope is *partial*. Coordination is *distributed*, *cooperative* and *partial*. The decision process is *dynamic* as it takes the actual load of neighbors into account. The algorithm is *sender-initiated* since nodes inform their neighbors about load changes and may receive entities if the load of neighbors is higher. The algorithm is not changed during runtime, therefore adaptivity is *fixed*. The cost sensitivity of the algorithm depends on the weights of the individual forces. Consequently, the are *multiple* possibilities for cost sensitivity. There is no explicit limit for the number of task migrations. Stability control is represented by the weighting of the "damping force" that is composed of a "frictional force" and a constant that defines the maximum number of migrations permitted for each task. Because if this constant stability control is classified as *guaranteed*.

The algorithm is based on a *multi-dimensional* index (Table 2.12 on page 58) that considers *CPU* and communication (*NET*). Aggregation can be classified as *multiple* and *hybrid*, because the load state depends not only on the actual load, but also on the resource requirements of the actually assigned tasks. Load value composition is based on *complex formulas* that are derived from the corresponding physical model. Weights corresponds to *system-characteristics* such as processor speeds and *pre-runtime application information* such as the number of instructions of a task. Index measurement is performed *implicit* when state changes occur. Index propagation is *adaptive*, because load values are only exchanged when load changes occur. The model is *fixed* since it is not changed during computation.

2.5.2 Domain Specific Load Distribution

Systems using domain specific load distribution consider domain specific semantics for load distribution decisions. For example, subproblem priorities may be determined according to the domain specific "importance" of a certain subproblem.

	EAGER et al.	FERGUSON et al.	HEISS et al.
Objectives			
domain specific	no	no	no
intent	system	system	both
function	assignment	assignment	assignment
Integration Level			
monitoring	system	system	system
strategy	system	system	system
mechanisms	system	system	system
compiler support	none	none	none
Structure			
problem structure	indep.	loc. dep.	loc. dep.
subproblems	processes	computation	computation
workspaces	-	processes	processes
targets	nodes	nodes	nodes

Table 2.10: General Classification of EAGER et al., FERGUSON et al., and HEISS et al.

	EAGER et al.	FERGUSON et al.	HEISS et al.
System Model			
Model Flavour	comb.	microecon.	physical
Target Top.	hom., full	het., graph	het., graph
Entity Top.	n.-i. entities	n.-i. entities	mesh
Transfer Model			
Transfer Space	systemwide	neighbor	neighbor
Transfer Policy	non-preemptive	preemptive	preemptive
Information Exchange			
Information Space	systemwide	neighbor	neighbor
Information Scope	partial	partial	partial
Coordination			
Decision Structure	distributed	distributed	distributed
Decision Mode	cooperative	competitive	cooperative
Participation	partial	partial	partial
Algorithm			
Decision process	dynamic	dynamic	dynamic
Initiation	sender	receiver	sender
Adaptivity	fixed	fixed	fixed
Cost Sensitivity	partial	high	mult.
Stability control	guaranteed	partial	guaranteed

Table 2.11: Strategy classification: EAGER et al., FERGUSON et al., and HEISS et al.

	EAGER et al.	FERGUSON et al.	HEISS et al.
Load Index Properties			
index dimension	single-dim.	multi-dim.	multi-dim.
index type	CPU	CPU, NET	CPU, NET
aggregation time	instantaneous	comprehensive	multiple
aggregation space	system related	hybrid	hybrid
Load Value Composition			
index comb.	-	lin. comb.	complex
weight sel.	-	sys., pre-run.	sys., pre-run.
Model Usage Policies			
index measurement	implicite	implicite	implicite
index propagation	on demand	adaptive	adaptive
model adaptivity	fixed	fixed	fixed

Table 2.12: Load Model Classification of EAGER et al., FERGUSON et al., and HEISS et al.

"Analysis of Dynamic Load Balancing Strategies for Parallel Shared Nothing Database Systems" by RAHM and MAREK [13]

RAHM et al. describe different strategies for dynamically managing the degree of parallelism and the assignment of subqueries in a parallel shared nothing database system. In the following, we always refer to the "DJP+ALUP" strategy. The DJP strategy performs dynamic adaption of the degree of join parallelism according to the actual multi-user degree. The ALUP strategy selects the least utilized processors as targets for processing subqueries. The "DJP+ALUP" is a combination of the DJP and the ALUP strategy.

The work of RAHM et al. is obviously *domain specific* (Table 2.13 on page 62). The intent is load distribution for a single application, namely a parallel database system, i.e. the intent is *application exclusive*. The function of load distribution is to manage *both* aspects, namely partitioning by adapting the degree of parallelism as well as assigning subproblems. Assignment decisions are based on application specific information and actual load information about nodes, therefore the approach requires *application-integrated* and *system-integrated monitoring*. Strategy and mechanisms are implemented within the *application*. The application structure consists of *independent* "subqueries" that are spread over the nodes. RAHM et al. just describe simulation results and not a concrete implementation. Assuming that subqueries are implemented as processes, we can classify the placement mechanism as assignment of processes to nodes.

The system model (Table 2.14 on page 63) can be classified as *combinatorial*. It consists of full-connected, homogeneous nodes (called "processing elements") that execute independent tasks ("subqueries"). Tasks may be distributed *systemwide*, assignment is *non-preemptive*. A central information base is periodically informed about the current load of all nodes. Therefore the information space is *central* and information scope is *complete*. Each query is initially assigned to an arbitrary node, that is thereupon responsible for the assignment of the subqueries of the query. In the following this node is called manager for that query. Ccoordination is *distributed*, as each query manager decides about the assignment of queries for which it is responsible. The central information base is used by the query managers for assignment decisions. Therefore, the decision mode is classified as *cooperative*. Participation is *partial*, since not all nodes are used for executing subqueries of a query. The decision process is obviously *dynamic*, mechanisms are initialized by the query managers, i.e. the *senders* of subqueries. The strategy adaptivity is *fixed*. Communication overhead is neglected, therefore the is *no cost sensitivity*. The system *requires no stability control*, because subqueries are only assigned once.

The load model (Table 2.15 on page 63) uses a *single-dimensional* load index cooresponding to the *CPU* usage. Aggregation time is *comprehensive*, aggregation space is *system related*. Nodes send there actual load *periodically* to the central information base. Index propagation is *on demand*, when a query manager ask the central information base for load values. Model adaptivity is *fixed*.

Migration mechanisms are not used.

"Data partitioning for multicomputer database systems: a cell-based approach" by HUA, LEE, and YOUNG [11]

HUA et al. (see also [12]) describe a method for dynamically partitioning database relations among a number of "processing nodes", consisting of a CPU and a local disk. The idea is to subdivide database relations into so called "cells" that are dynamically assigned to nodes. The goal is to partition the relations in order to keep the number of records stored on each node the same, regardless of their usage pattern.

HUA et al. present a method for *domain specific, application oriented* load distribution (Table 2.13 on page 62). Although the authors assume a possibility to dynamically split or merge cells, the presented algorithm is only responsible for non-preemptive *assignment* of a given number of cells to nodes, but not for partitioning relations into cells. Monitoring, strategy, and mechanisms are intended to be integrated into the *application*. Load is implied by accesses to records. Therefore, subproblems are represented by *accesses to application objects* (cells). Load distribution is realized by implementing *application objects* (cells) as workspaces that are preemptively assigned to *nodes*. Nodes represent computing units with a local disk.

The system model (Table 2.14 on page 63) has a *combinatorial* flavour. Nodes are assumed to be *homogeneous* and *full connected*. There is *no specific topology* among entities (application objects). Application objects can be transferred *systemwide*. The assignment of application objects is *preemptive*. HUA et al. assume a central instance that controls the assignment based on global information. Therefore, information exchange is *central* and *complete*, and the decision structure is *centralized, cooperative*, and uses *global participation*. The decision process is *dynamic*, because considers the actual number of records within each cell. HUA et al. make no statement which instance is responsible to initialize a re-distribution. The adaptivity is *fixed*. The cost sensitivity may be rated as *partial*, because the algorithm tries to minimize the number of transferred objects, but is does not explicitly consider system resource usage such as communication amount etc. Stability control is *not required*, because an object can be migrated only once within each redistribution phase.

The load model (Table 2.15 on page 63) uses a *single-dimensional* load index, namely the actual number of records within a cell. Therefore, the index type is *application defined*, aggregation time is *instantaneous*, and aggregation space is *application related*. There is no load value composition. Index measurement can be implemented implicitly just by using a counter for the number of records within a cell. Index propagation is *on demand* when a redistribution phase is started by the central load distribution instance. The model adaptivity is *fixed*.

Table 2.16 on page 64 shows the migration mechanism classification. Since migration is implemented within the application, it is obviously *not transparent*. Each migration step consists of moving one cell to another node, hence migration set size is *limited*. Migrations take place in a special "redistribution phase". Therefore, migration requests are performed *immediately*. Migration between heterogeneous nodes is not implemented. Note that this would be a non-trivial task because migrated data records are structured data types. *No pre transfer media* and *no compression* is used. Cells are *completely* transferred.

"Automatic Tuning of Data Placement and Load Balancing in Disk Arrays" by SCHEUER-MANN, WEIKUM, and ZABBACK [14]

SCHEUERMANN et al. (see also [16]) describe a file management algorithm (called "GDC" for "Generalized Disk Cooling") for dynamically assigning partitions of files to disks. Each file is partitioned into "striping units" that are used as distribution entity.

Regarding the general classification (Table 2.13 on page 62), the concept is obviously *domain specific*. Although the system is intended to be integrated into an operating system, it is classified as *application specific* since it does not provide general load distribution features for applications on top of the operating system, but deals only with the assignment of a specific class of operating system objects. Although [14] mentions heuristics for automatically determining striping units, we only consider the presented GDC algorithm that deals with *assignment* of given striping units. The entire file management system is regarded as an application, That is why monitoring, strategy, and mechanisms are classified as *application integrated*. The problem structure consists of *location dependent* subproblems (accesses to striping units): Data accesses are obviously bound to the respective striping unit since there is no data replication. Striping units (i.e. application objects) correspond to workspaces that are migrated between nodes (using a direct correspondence between disks and nodes).

The strategy uses a combinatorial model (Table 2.14 on page 63). Targets (here: disks) are assumed to be *homogeneous* and *full connected*. Entities (objects) are *non-interacting*. Objects can be *preemptively* transferred *systemwide*. Strategy decisions are based on *central* and *complete* information. The decisions are performed by a *central* component that *cooperates* with the individual disks. Only the sending and the receiving disk are involved in a migration, hence participation is *partial*. The decision process is *dynamic* as it is based on the actual access pattern to objects. Initiation of migrations is managed by the *central* decision component. Adaptivity is *fixed*. Cost sensitivity is *high* by introducing a "cost factor" into the decision whether a migration will improve performance. There is *no stability control*.

The load model (Table 2.15 on page 63) consists of a *single-dimensional* load index called the "heat" of a disk. The heat corresponds to the access intensity to a disk. It is related to *I/O* (secondary storage load). Aggregation time is *comprehensive* by observing heat within certain time spans. The system is regarded as an application, hence the aggregation space is *application related*. Although a single-dimensional load index is used, several values such as heat with respect to read or write operations are combined by *complex* formulas. *Pre-runtime application information* such as periodical heat patterns over time are used. Load (heat) measurement and propagation is initialized by the central manager and can be classified as *on demand*. The model adaptivity is *fixed*.

Regarding the system as an application, the migration classification is similar to that of HUA et al. (Table 2.16 on page 64). The sole difference is the initiation. If the strategy decides to migrate an object, the corresponding "action" is inserted into the working queue of the sending node. Because these migration actions have a lower priority than conventional file accesses, initiation is classified as *delayed*.

"Dynamic Load Balancing of Data Parallel Applications on a Distributed Network" by HAMDI and LEE [7]

HAMDI and LEE describe a load distribution method that is integrated into a parallel image processing program.

Table 2.13 shows the general classification. The load distribution method is *domain specific* as it is specific to the structure of image processing algorithms. The intent is to distribute workload among processes that run in a non-exclusively used multi-user workstation network. Therefore the intent is *adaptive application oriented* load distribution. The function is *assignment*, since the granularity of the partitioning (here: number of processes) is not changed. Monitoring, strategy, and mechanisms are implemented within the *application*. The problem structure is *location dependent*: Subproblems (*data accesses*) have to be performed by processes holding the corresponding data, i.e. workspaces correspond to *data* and *processes* are assignment targets for workspaces.

The strategy (Table 2.14 on page 63) uses a *combinatorial* system model consisting of full connected, generally heterogeneous targets (nodes). The entity topology is a *ring* as each process communicates only with its left and right neighbors according to the stripwise decomposition of the image. Data (workspaces) is migrated only between *neighbors*, i.e. data assignment is *preemptive*. The decision structure is *centralized*, as new assignments are computed by a central instance that uses *central* and *complete* load information. Nevertheless, the individual processes realize a new assignment only if they they are "underloaded" and require data from their neighbors. Therefore, the decision mode is *cooperative* and *partial*. The algorithm is *dynamic* since it can react on internal load changes as well as to load changes due to other users in the system. Initiation of mechanisms is by *receivers* of data (see above). Adaptivity is *fixed*. The cost sensitivity may be classified as *partial*, because actual communication overhead is not directly considered. There is no stability control.

The load model (Table 2.15 on page 63) uses a *single-dimensional* load index corresponding to the number of "pixels" that have been processed between the arrival of two consecutive "load request messages". The index type is *CPU*. Aggregation time is *comprehensive* since a time interval is considered. Aggregation space is the *application*, system load is not considered directly. Index measurement is *implicit* because the number of processed pixels is represented by the loop counters of the image processing algorithm. Index propagation is *on demand*, i.e. when a corresponding request from the central decision instance is received by a process. Model adaptivity is *fixed*.

The migration mechanism (Table 2.16 on page 64) is similar to that of SCHEUERMANN et al. The single difference is the *unlimited* migration set size, since an arbitrary number of workspaces (data columns) can be migrated within one migration step.

	RAHM e.a.	HUA e.a.	SCHEUERMANN e.a.	HAMDI e.a.
Objectives				
domain specific	yes	yes	yes	yes
intent	appl. excl.	appl. excl.	appl. excl.	adaptive
function	both	assignment	assignment	assignment
Integration Level				
monitoring	appl. and system	appl.	appl.	appl.
strategy	appl.	appl.	appl.	appl.
mechanisms	appl.	appl.	appl.	appl.
compiler support	none	none	none	none
Structure				
problem structure	indep.	loc. dep.	loc. dep.	loc. dep.
subproblems	processes	obj. accesses	object accesses	data accesses
workspaces	-	objects	objects	data
targets	nodes	nodes	nodes	processes

Table 2.13: General Classification of RAHM et al., HUA et al., HEISS et al., and HAMDI et al.

2.5.3   Run-Time Environments for Load Distribution

"Efficient Parallel Computing in Distributed Workstation Environments" by CAP and STRUMPEN [4]

The PARFORM platform (see also [3; 15]) has been designed as a complete environment for parallel computing in workstation networks. Load distribution is one important aspect in the design of this system.

Table 2.17 on page 66 shows the general classification. The PARFORM is *not domain specific* as it can be applied to a large range of application domains. The intent is *adaptive* load distribution for parallel applications. Dynamic partitioning is not supported, i.e. the function of the PARFORM is *assignment*. Monitoring, strategy, and mechanisms are provided by the PARFORM runtime system. Generally, PARFORM applications are *location dependent*. *Data accesses* (representing subproblems) are performed on local *data* (representing workspaces). The application data is assigned to application *processes*.

CAP et al. give no detailed description of the general load distribution strategy and load indices that are used by the platform. This section classifies the application example (PDE solver) mentioned in [4], using dynamic load distribution (the paper describes also variants with static homogeneous and static heterogeneous partitioning). Regarding the strategy (Table 2.18 on page 67), the system model is *combinatorial* using a virtually *full* connected set of *heterogeneous* workstations. The entity topology is a *ring*, i.e. each process communicates only with two neighbors. Assignment is *preemptive*. Data is migrated between *neighbored* processes according to the ring-topology. Load information is only exchanged between *adjoining* processes, information scope is partial. Two adjoining processes decide independently from other processes whether they exchange data structures. Therefore, coordination is *distributed*, *cooperative*, and *partial*. The decision process is *dynamic* since actual system load is considered. Migration of data structures is

	RAHM e.a.	HUA e.a.	SCHEUERMANN e.a.	HAMDI e.a.
System Model				
Model Flavour	comb.	comb.	comb.	comb.
Target Top.	hom., full	hom., full	hom., full	het., full
Entity Top.	n.-i. entities	n.-i. entities	n.-i. entities	ring
Transfer Model				
Transfer Space	systemwide	systemwide	systemwide	neighbor
Transfer Policy	non-preemptive	preemptive	preemptive	preemptive
Information Exchange				
Information Space	systemwide	central	central	central
Information Scope	complete	complete	complete	complete
Coordination				
Decision Structure	distributed	centralized	centralized	centralized
Decision Mode	autonomous	cooperative	cooperative	cooperative
Participation	partial	global	partial	partial
Algorithm				
Decision process	dynamic	dynamic	dynamic	dynamic
Initiation	sender	-	central	receiver
Adaptivity	fixed	fixed	fixed	fixed
Cost Sensitivity	none	partial	high	partial
Stability control	not required	none	none	none

Table 2.14: Strategy classification: RAHM et al., HUA et al., SCHEUERMANN et al., and HAMDI et al.

	RAHM e.a.	HUA e.a.	SCHEUERMANN e.a.	HAMDI e.a.
Load Index Properties				
index dimension	single-dim.	single-dim.	single-dim.	single-dim.
index type	CPU	application	I/O	CPU
aggregation time	comprehensive	instantaneous	comprehensive	compreh.
aggregation space	system related	application	application	application
Load Value Composition				
index comb.	-	-	complex	-
weight sel.	-	-	pre-run.	-
Model Usage Policies				
index measurement	periodic	implicite	on demand	implicit
index propagation	on demand	on demand	on demand	on demand
model adaptivity	fixed	fixed	fixed	fixed

Table 2.15: Load Models of RAHM et al., HUA et al., SCHEUERMANN et al., and HAMDI et al.

	HUA e.a.	SCHEUERMANN e.a.	HAMDI e.a.
Mig. Transp.	no	no	no
Code Transp.	no	no	no
Set Size	limited	limited	unlimited
Initiation	immediate	delayed	delayed
Res. Dep.			
Heterogeneity	no	no	no
Pre Transfer Media	-	-	-
Compression	no	no	no
Transfer Pol.	complete	complete	complete

Table 2.16: Migration mechanism classification: HUA et al., SCHEUERMANN et al., and HAMDI et al.

initiated by the heavier loaded node, i.e. by the *sender*. Adaptivity is fixed. Cost sensitivity is *partial* because the actual communication amount is not considered. There is *no stability control*.

The load model (Table 2.19 on page 67) is *multi-dimensional* since idle time percentages as well as available performance (measured by running a small test load) are used. The only index type is *CPU*. Aggregation time is *comprehensive*, because idle time percentages and available performance represent the load over a certain time span. Aggregation space is obviously *system related*. Index combination is implemented as *boolean selection*, i.e. either idle time percentage or available performance is used. The latter is used for initial allocation and the former is used for migration. Index measurement is *periodically fixed*. Index propagation is adaptive: If a process sends a message, its actual load value is appended to that message. The index combination is changed from available performance in the initial phase to idle time percentage afterwards. Nevertheless, we classify model adaptivity as *fixed* as the change is independent from the system behavior.

Migrations are *not obvious* to the application (Table 2.20 on page 68). Nevertheless, applications have to be adapted to the PARFORM platform, therefore use of migration mechanisms is obviously *not source code transparent*. Migration set size is *limited*, since only one entity (column of a matrix) is migrated in one step. Initiation is *immediate*, because the heavily loaded process immediately sends an entity to a less loaded neighbor. Heterogeneity is possible since only data is migrated. No compression and no pre transfer media is used. Entities (matrix columns) are migrated *completely*.

"Performance of dynamic load balancing algorithms for unstructured mesh calculations" by WILLIAMS [17]

The DIME environment (see also [18]) provides functions for assigning irregular meshes tp processes in parallel systems. The application attaches computing code for elements and nodes of the mesh whereas DIME controls assignment, communication and synchronization. In the following we classify the DIME environment using the ORB (Orthogonal Recursive Bisection) method as described in [18].

Table 2.17 on page 66 shows the general classification. DIME is *domain specific* as it is designed for computations on mesh data structures. The intent is to *assign* approximately the same amount of workload to each node in the system. Hence, the intent is *application exclusive* load distribution. Monitoring, strategy and mechanisms are completely integrated into the *runtime system*. Regarding the problem structure, the application consists of *location dependent* subproblems (*data accesses*). Workspaces correspond to *data* which is used in computations. The *nodes* represent targets for data-assignment.

The load distribution strategy of DIME is classified in Table 2.18 on page 67. The model flavour is combinatorial. Targets (nodes) are assumed to be *fully connected* and *homogeneous*. The entity topology is a *mesh*. Transfer space is *systemwide* since ORB does not consider the actual distribution for computing a new distribution. Transfer policy is *preemptive* assignment of data. ORB uses a *central* and *complete* information exchange. Coordination is *centralized, cooperative*, and *global* since redistribution is performed in a separate phase. The algorithm is *dynamic* as it considers the actual number of data elements within a region. This number may dynamically change due to refinements of the mesh. Initiation of mechanisms is performed after each refinement, hence it can be classified as *central*. Adaptivity is *fixed*. There is *no cost sensitivity*, since ORB does neither consider actual assignments for computing a new assignment, nor communication costs for migration. Stability control is *not required*, because data can be migrated only once after each refinement step.

The load model (Table 2.19 on page 67) of DIME is *single-dimensional* as it just considers the actual number of data elements within a region. The computational requirements of a process are assumed to be proportional to the number of data elements assigned to that process. Therefore, the index type is *CPU*. Aggregation time is *instantaneous* because ORB considers the actual number of data elements, Aggregation space is obviously *application related*. There is no load value composition. Index measurement is *on demand*: If the ORB is started, DIME has to compute the number of mesh elements within a certain region. Index propagation, i.e. performing an ORB phase, is *adaptive* since it is started after each refinement step. Model adaptivity is *fixed*.

The migration mechanism (Table 2.20 on page 68) is *transparent with respect to performing migrations*, but *not source code transparent*. Code has to be modified for using DIME. Migration set size is *unlimited*. Initiation of migrations can be classified as *immediate*, since it is performed in a separate phase. *Heterogeneity is not supported. No pre transfer media* and *no compression* is used. The data is migrated *completely*.

"Dynamic Balancing Complex Workload in Workstation Networks - Challenge, Concepts and Experiences" by BECKER [2]

BECKER (see also [1]) describes the HiCon model. HiCon assigns processes non-preemptively to heterogeneous nodes, considering load induced by other users. Furthermore, data structures are assigned *preemptively* to processes. In contrast to the previously regarded systems, HiCon integrates several different strategies.

In general, HiCon is *not domain-specific* (Table 2.17). It is intended for distribution within parallel applications as well as distribution among multiple parallel applications. The function is *assignment* as well as *partitioning*. Monitoring is integrated into the runtime system and uses additional operating system specific information. Strategy and mechanisms are integrated into the HiCon runtime system. The structure of HiCon applications is generally *location dependent*. *Data accesses* (subproblems) are performed on *data* (workspaces) that is assigned preemptively to the processes of the parallel application. Furthermore, HiCon assigns processes dynamically but non-preemptively to nodes.

There are several different load distribution strategies for *HiCon*. All strategies use a combinatorial system model. The target ("processor") topology consists of *heterogeneous clustered* nodes. The application can define an "affinity graph", therefore the entity topology is generally a *mesh*. HiCon assigns two types of entities: It assigns processes *non-preemptively* to nodes as well as data *preemptively* to processes. Assignment decisions are based on a hierarchical structure: Within a cluster, load distribution is performed by a central cluster manager whereas load distribution among clusters is performed by a distributed protocol among cluster managers. Therefore, information exchange may be classified as *systemwide* transfer/information space and *partial* information scope. Coordination is *hierarchical, cooperative* and *partial*. The decision process is *dynamic* since it considers several actual parameters. Initiation of load distribution may be classified as central as it is performed by the cluster manager. There is *no adaptivity*. Depending on the used strategy, there are *multiple possibilities* for using cost sensitivity. Stability control is not mentioned.

HiCon provides several different load models (Table 2.19) that are multi-dimensional. All index types mentioned in section 2.3 may be used. Aggregation space is *hybrid*. There are several possibilities for index combinations. Weight selection may be based on system characteristics as well as pre-runtime application information such as communication affinity. There are *multiple* possibilities for index measurement. Index propagation is *on demand*, when a cluster manager collects information. The load model may use *adaptive weighting*, such as dynamic adjustment of the relationship between overhead of load distribution and profit by balanced workload distribution.

BECKER gives no detailed description of migration mechanisms.

	CAP et al.	WILLIAMS	BECKER
Objectives			
domain specific	no	yes	no
intent	adaptive	application	both
function	assignment	assignment	both
Integration Level			
monitoring	rts	rts	rts/system
strategy	rts	rts	rts
mechanisms	rts	rts	rts
compiler support	none	none	none
Structure			
problem structure	loc. dep.	loc. dep.	loc. dep.
subproblems	data accesses	data accesses	data acc./processes
workspaces	data	data	data
targets	processes	nodes	processes/nodes

Table 2.17: General Classification of CAP et al., WILLIAMS, and BECKER

	CAP et al.	WILLIAMS	BECKER
System Model			
Model Flavour	comb.	comb.	comb.
Target Top.	het., full	hom., full	het., clustered
Entity Top.	ring	mesh	mesh
Transfer Model			
Transfer Space	neighbor	systemwide	systemwide
Transfer Policy	preemptive	preemptive	preemp. data non-preemp. proc.
Information Exchange			
Information Space	neighbor	central	systemwide
Information Scope	partial	complete	partial
Coordination			
Decision Structure	distributed	centralized	hierarchical
Decision Mode	cooperative	cooperative	cooperative
Participation	partial	global	partial
Algorithm			
Decision process	dynamic	dynamic	dynamic
Initiation	sender	central	central
Adaptivity	fixed	fixed	fixed
Cost Sensitivity	partial	none	mult.
Stability control	none	not required	none

Table 2.18: Strategy classification: CAP et al., WILLIAMS, and BECKER

	CAP et al.	WILLIAMS	BECKER
Load Index Properties			
index dimension	multi-dim.	single-dim.	multi-dim.
index type	CPU	CPU	multiple
aggregation time	comprehensive	instantaneous	multiple
aggregation space	system related	application	hybrid
Load Value Composition			
index comb.	-	-	multiple
weight sel.	-	-	sys., pre-run.
Model Usage Policies			
index measurement	periodically	on demand	multiple
index propagation	adaptive	adaptive	on demand
model adaptivity	fixed	fixed	ad. weights

Table 2.19: Load Model Classification of CAP et al., WILLIAMS, and BECKER

	CAP et al.	WILLIAMS
Mig. Transp.	yes	yes
Code Transp.	no	no
Set Size	limited	unlimited
Initiation	immediate	immediate
Res. Dep.		
Heterogeneity	yes	no
Pre Transfer Media	-	-
Compression	no	no
Transfer Pol.	complete	complete

Table 2.20: Migration mechanism classification: CAP et al. and WILLIAMS

References

[1] W. BECKER. Das HiCon-Modell: Dynamische Lastverteilung für datenintensive Anwendungen in Workstation-Netzen. Technical Report 1994/4, Universität Stuttgart, Fakultät für Informatik, 1994.

[2] W. BECKER. Dynamic Balancing Complex Workload in Workstation Networks - Challenge, Concepts and Experiences. In *High-Performance Computing and Networking*, pages 407–412, Milan, Italy, May 1995. Springer, LNCS 919.

[3] C. CAP and V. STRUMPEN. THE PARFORM – A High Performance Platform for Parallel Computing in a Distributed Workstation Environment. Technical Report 1992/7, Universität Zürich, Institut für Informatik, 1992.

[4] C. CAP and V. STRUMPEN. Efficient Parallel Computing in Distributed Workstation Environments. *Parallel Computing*, 19(11):1221–1234, 1993.

[5] D. L. EAGER, E. D. LAZOWSKA and J. ZAHORJAN. Adaptive Load Sharing in Homogeneous Distributed Systems. *IEEE Transactions on Software Engineering*, 12(5):662–675, 1986.

[6] D. FERGUSON, Y. YEMINI and C. NIKOLAOU. Microeconomic Algorithms for Load Balancing in Distributed Computer Systems. In *Proceedings of the 8th International Conference on Distributed Computing Systems, San Jose, California*, pages 491–499, 1988.

[7] M. HAMDI and C.-K. LEE. Dynamic Load Balancing of Data Parallel Applications on a Distributed Network. In *International Conference on Supercomputing*, Barcelona, Spain, 1995.

[8] H.-U. HEISS and M. SCHMITZ. Distributed Load Balancing Using a Physical Analogy. Technical Report 5/92, Universität Karlsruhe, Fakultät f'ur Informatik, May 1992.

[9] H.-U. HEISS and M. SCHMITZ. Decentralized Dynamic Load Balancing: The Particles Approach. In L. GRÜN, R. ONVURAL and E. GELENBE, editors, *In Proc. of the 8th Int. Symp. on Computer and Information Sciences*, Instanbul, Turkey, 1993.

[10] H.-U. HEISS and M. SCHMITZ. Decentralized Dynamic Load Balancing: The Particles Approach. *Information Sciences*, 84(1+2):115–128, 1995.

[11] K. A. HUA, C. LEE and H. C. YOUNG. Data partitioning for multicomputer database systems: a cell-based approach. *Information Systems*, 18(5):329–342, 1993.

[12] K. A. HUA and H. C. YOUNG. A Cell–Based Data Partitioning Strategy for Efficient Load Balancing in A Distributed Memory Multicomputer Database System. Technical Report RJ 8041, IBM Research Report, 1991.

[13] E. RAHM and R. MAREK. Analysis of Dynamic Load Balancing Strategies for Parallel Shared Nothing Database Systems. In *Proceedings of the 19th International Conference on Very Large Data Bases, Dublin, Ireland*, 1993.

[14] P. SCHEUERMANN, G. WEIKUM and P. ZABBACK. Automatic Tuning of Data Placement and Load Balancing in Disk Arrays. Technical Report 175, ETH Zürich, April 1992.

[15] V. STRUMPEN. Parallel Molecular Sequence Analysis on Workstations in the Internet. Technical Report 1993/28, Universität Zürich, Institut für Informatik, 1993.

[16] G. WEIKUM, P. ZABBACK and P. SCHEUERMANN. Dynamic File Allocation in Disk Arrays. In *Proceedings of the 1991 ACM SIGMOD International Conference on Management of Data*, pages 406–415, Denver, Colorado, 1991.

[17] R. WILLIAMS. Performance of dynamic load balancing algorithms for unstructured mesh calculations. *Concurrency: Practice and Experience*, 3(5):457–481, 1991.

[18] R. WILLIAMS. DIME: Portable Software for Irregular Meshes for Parallel or Sequential Computers. In *COMPCON Spring '91*, pages 68–72, San Francisco, California, February 1991.

3 Systems

The previous chapter has introduced classification schemes for all aspects of load distribution. This chapter now provides an overview on different load distribution systems that have been developed in our research institute and applies the previously defined taxonomy to the systems.

The first section describes a migration mechanism that allows to migrate entire processes during runtime. In section 3.2 a load distribution system based on a distributed thread kernel is presented. After that section 3.3 explains how a microeconomic approach can be used to address the problem of load distribution. Finally, section 3.4 deals with a hybrid between system based and application based load distribution system.

3.1 Consistent Checkpointing

by Georg Stellner

CoCheck[1] is a run-time environment for parallel applications. CoCheck allows both the creation of checkpoints and the migration of processes of parallel applications on networks of workstations. Initially, CoCheck extended PVM [3], so that PVM applications could be started under the control of a resource management system [11]. In that case, CoCheck was used to create checkpoints in order to provide global scheduling of parallel applications. Although process migration was already supported, its performance needed further improvement. Consequently, the focus of the research was set on performance improvements of checkpointing and particularly process migration. This could be achieved by transferring the checkpoints directly over TCP network connections [10]. As the next step, CoCheck was implemented to support the proposed MPI [8] message passing standard. Therefore, the protocol was integrated with tuMPI[2] which is an implementation of the MPI standard definition [12]. For that implementation of CoCheck a decision component was added that performs automatic load distribution by process migration. Recent research focused on formal specifications and correctness proofs of CoCheck [7]

The remainder of this section is organized as follows. Firstly, the CoCheck protocol will be introduced. Then its implementation for tuMPI will be explained. This is followed by a presentation of the automatic load distribution system which has been added. After that performance results of the load distribution system will be discussed. Finally, a conclusion and an outlook on future work is given.

3.1.1 Outline of the Algorithm of CoCheck

In addition to designing a checkpointing facility for parallel applications on networks of workstations, a further goal of CoCheck was to achieve portability. Apart from supporting various workstation platforms the basic approach should be adaptable to different message passing libraries. First, because checkpointing a running process is inherently machine dependent and difficult to implement, CoCheck should be layered on top of a portable single process checkpointing

---

[1] Consistent **Checkpoints**

[2] Technische Universität München Message Passing Interface

mechanism such as [4; 6]. Second, an implementation which is integrated into the message passing systems reduces the suitability for other message passing libraries. Consequently, CoCheck should be implemented on top of the parallel environment rather than inside. Figure 3.1 shows how CoCheck is integrated in the existing software structure of a network of workstations.

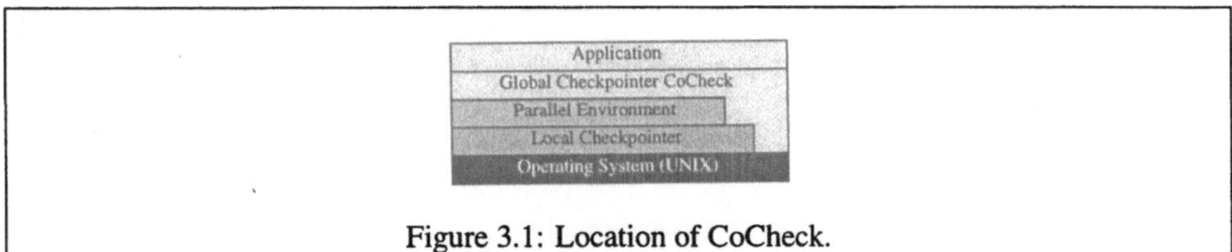

Figure 3.1: Location of CoCheck.

An important issue from the applications' point of view is transparency of checkpointing, i.e. the application need not to be aware that either an application has been restarted from a checkpoint or some of its processes have been migrated. Particularly, it is inevitable that the application can use the same addresses for the processes throughout its lifetime. Hence, we introduced *virtual addresses*. The initial address which is associated with the process when it is started is used as the address in the application, whereas CoCheck actually uses the current addresses of the processes. Therefore, a mapping from original to current addresses has to be maintained. As we do not make any assumptions about the availability of additional daemon processes, we decided that each process of the parallel application has its own copy of the mapping table. Upon migration these tables have to be updated. To hide all the additional functionality like virtual addresses or message buffering (explained below), we are providing wrapper functions for all calls of the message passing library. When the user links his application these wrappers are used instead of the original functions. The wrappers in turn call the original functions to perform the corresponding service. The wrapper functions add additional functionality if necessary. In the framework of MPI the profiling interface [8] can be used, whereas for other message passing libraries the names of the calls are directly renamed within the library by a tool to avoid source code modifications.

Messages which are being in transit when a checkpoint should be taken or a process should be migrated are a major problem. Messages can be found in various places such as the physical network, operating system buffers or buffers of the message passing library (which in turn can be located in an additional daemon or locally within the processes or both). Moreover, it is difficult to access these locations and to capture the state of the network to get a consistent global state of the application. In addition, accessing the messages in these places would also violate the requirement that the implementation of CoCheck sits on top of the message passing library and not underneath it. If the message passing library preserves the sending sequence of the messages when they are received, the following solution can be applied to get around this problem.

CoCheck forces a situation in which no more messages are in transit. Therefore, special messages, so called "ready messages" (RM), are used to clear the potential communication channels between the processes. Each process in the application is notified that a checkpoint has to be taken or a set of processes has to be migrated. It is important that this notification is not processed while a send operation is in progress. Otherwise, the consistency of the global state could be violated. Consequently, the send operation has to be atomic, which is also achieved in the wrapper functions for the send calls. Each process continues with sending a RM to all the other processes. Besides, each process receives any incoming message. If a process receives a RM from a certain process,

it can conclude that there will not follow further messages from that process, i.e. that this channel
has been cleared. However, if the message is not a RM, it is stored in a reserved buffer area in
the data segment of the process. This receive operation anticipates the receive operation of the
application.

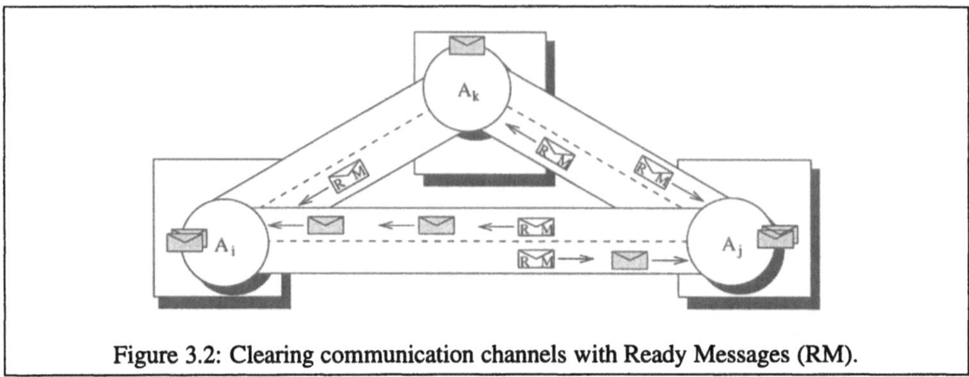

Figure 3.2: Clearing communication channels with Ready Messages (RM).

Figure 3.2 shows the effect of the RMs: all messages in transit before the RM are delivered to
the process and buffered there. When a process has collected the RMs from all other processes,
it can be sure that there are no further messages on the network with its address as destination.
All messages that were in transit are safely stored in the buffer and are available for later access.
Hence, the process can be seen as an isolated single process for which it is easy to determine
its current state and create a checkpoint with the single process checkpointer. As no messages
are being transmitted at checkpoint time, there is also no need to forward messages after restart.
Instead, the receive calls of the processes will first try to extract matching messages from the
buffers in the data segments.

After restart the address mapping tables have to be updated with the new current addresses
of the processes. Afterwards the processes continue to execute the user programm where they
were interrupted by the notification. If they are executing a receive operation, the receive first has
to check the buffer for a matching message. If such a message is available, it is retrieved from
the buffer and returned. Otherwise, a real receive operation intercepts the next matching message
from the network.

The approach described so far assumes one special process which serves as a coordinator of
the activities. As this can either be a special system daemon being responsible for resource management or dynamic load distribution, or a user command triggering checkpointing or migration,
this is no restriction. The outline of the algorithm can be summarized as follows:

1. **central instance:** send notification to all processes
2. **each process:** send RM to all other processes
3. **each process:**
    - receive incoming messages until all RMs have been received
    - store user messages in buffer and count RM
4. **each process:** disconnect from parallel environment
5. **each process:** create checkpoint or migrate

The restart can be summarized as follows:

1. **central instance:** restart all processes
2. **each process:** reconnect to parallel environment
3. **each process:** send new address to central instance
4. **central instance:** collect new addresses and distribute the new mapping table to all processes
5. **each process:** resume execution (using the new addresses and the messages in the buffer area)

3.1.2 Implementation of CoCheck for MPI

MPI is the proposed standard for message passing programming. Therefore, the portability of the CoCheck approach should be demonstrated with an implementation of the MPI standard. At that time it turned out that all publicly available MPI implementations were based on another message passing library or were part of a larger environment. Hence, we decided to first bring up our own native MPI implementation. This could be done in reasonable amount of time as much code could be borrowed from MPICH [1] for the higher level MPI functionality and NXLib [9] for the communication part.

3.1.2.1 Implementation of tuMPI

Primarily tuMPI should be a research environment to investigate dynamic load distribution based on process migration. Other goals like efficiency or support for heterogeneity were not as important or left open for future work.

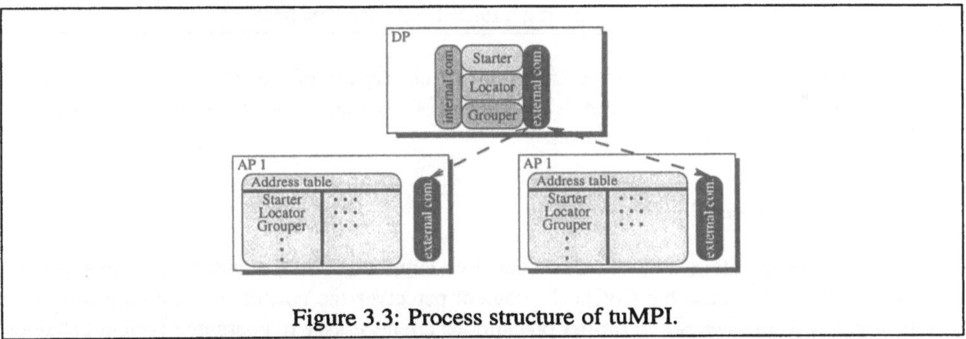

Figure 3.3: Process structure of tuMPI.

The process structure of a small sample tuMPI application is shown in Figure 3.3. It consists of two application processes (AP) and a daemon (DP) which comprises three major components: the starter, the locator and the grouper. The starter is used to start APs, whereas the locator is an address server. It knows about the exact whereabouts of all APs. Finally, the grouper is used to manage MPI groups and contexts. Every AP has a local address table which is filled when the application is performing send and receive operations. TCP connections to other APs are set up on demand with the assistance of the locator. The mapping of MPI addresses to TCP addresses is entered in the mapping table of the two APs that start to communicate and in the mapping table of the locator. With this scheme all APs that need to exchange messages are fully connected via TCP sockets.

3.1.2.2   CoCheck for tuMPI

A basic property of MPI, namely that multiple calls to the enrollment function `MPI_Init` are undefined [8], made it impossible to go through the cycle of disconnecting and reconnecting in the checkpoint protocol. At this point, CoCheck had to be integrated into the sources of tuMPI. The same consequence will also hold for other MPI implementations in concert with CoCheck. Finally, as the tuMPI sources had to be modified anyway, the central instance of CoCheck was integrated as an additional component in the DP.

The algorithm to create a consistent state was optimized for process migration in the following way. On the destination node a skeleton process (SP) is started. It delivers a notification to the process that ought to be migrated. The migrating process (MP) sends a RM to all its communication partners after it has intercepted the notification. All current communication partners can be determined from the local mapping table. All APs which have received a RM in turn send a RM back to MP. Again, until MP has received all RMs, incoming users messages are buffered. When MP has collected all RMs, it uses the Condor checkpointer to migrate to the new node by transferring local state over a TCP connection to SP. Figure 3.4 summarizes the migration of MP.

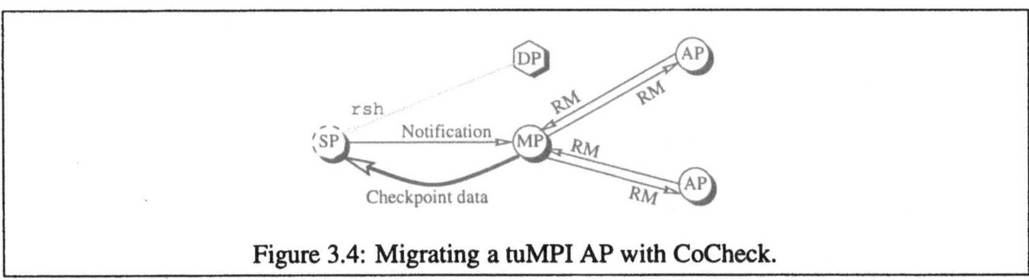

Figure 3.4: Migrating a tuMPI AP with CoCheck.

A slightly modified protocol can be used to take a checkpoint of the complete application. In that case, the notification is sent to all APs and instead of transfering the checkpoint directly to SP, each process writes its checkpoint on disk.

3.1.2.3   Further Assumptions

The current implementation assumes that collective communication is based on point-to-point communication. In this case the CoCheck protocol perceives the collective communication just as code written by the user and no further precautions are necessary to guarantee correct behavior after restart.

Currently the semantics of synchronous sends may be changed by a checkpoint: an anticipated receive may trigger the return of the send instead of a receive of the application. In the future, this can be changed by a slight modification of the wrappers, where an additional synchronization is introduced at the end of the synchronous send and receive operation.

3.1.3   Automatic Load Distribution with CoCheck for tuMPI

The automatic load balancer which has been implemented for tuMPI consists of three components: a load measure component, a decision component and a migration component. The latter is provided by the CoCheck extension for tuMPI and has been discussed in previous sections. The remaining two components will be introduced now. Both have been added as an additional part to the tuMPI DP.

3.1.3.1 Load Measurement

In accordance with literature where simple load indices and strategies achieved good results [2] the automatic load balancer integrated into tuMPI also uses a simple load index. It is based on the average length of the ready queue on a node during the last minute ($avenrun$). This value is available on every UNIX system and can easily be determined. With the `rup` command the value of $avenrun$ can also be determined on remote nodes, so that there is no need for additional monitor processes on the nodes which execute processes of parallel applications.

At startup time of an application the user can specify two additional parameters which configure the intervals at which the load should be determined. The first parameter specifies the time between two successive measurements when no migration was necessary. The second parameter determines the time at which the next measurement is performed when a process was migrated. This facility has been introduced to allow more time after a migration for the load values to stabilize. This avoids too many migrations to a machine before the additional load of a process has shown its effects also in the $averun$ value. Since the complete parameters with which an application is started are given to the call of `MPI_Init` those two parameters are removed from the parameter list during that function. Hence, the application does not need to be changed to handle the additional parameters: `MPI_Init` provides the desired transparency.

3.1.3.2 Decision Component

Although the migration of processes with CoCheck is only possible between migration compatible (homogeneous) machines[3] the decision component supports heterogeneous networks of workstations. Therefore the network is divided in homogeneous sub-clusters and the decision component tries to evenly distribute the load within each sub-cluster. As even within homogeneous sub-clusters the potential computational power of the machines can vary due to different clock frequencies, main memory capacity, etc. the above mentioned $avenrun$ value of each machine is normalized with a machine specific architectural constant α. In the current implementation the user is responsible to assign this constant to each machine in the mapping table that specifies the cluster and that is processed by tuMPI during startup. The load index that is used in the decision component is provided in (3.1).

$$load \ = \ \frac{1 + avenrun}{\alpha} \tag{3.1}$$

Increasing the $avenrun$ value by one in the denominator guarantees that in case of unloaded machines ($avenrun = 0$) machines with different speeds can actually be distinguished. Hence, the decision component can choose the potentially faster machine in case a destination machine for a migration has to be determined.

---

[3]Migration between binary compatible machines is not always possible due to different run-time properties of processes. In the case of Sun machines the binaries can be run on any machine, but it is not possible to migrate a process from a sun4m to a sun4c implementation of the SPARC specification due to a different run-time stack. Hence the requirement of binary compatible machines is not sufficient.

The evaluation phase begins after the load on all machines has been determined. Therefore the normalized load indices of all nodes are calculated using (3.1). After that the difference between the highest and lowest normalized value is calculated. If this difference exceeds a specified imbalance value a migration is considered. Otherwise no migration will be performed and the next measurement starts after the waiting time that has been specified (c.f. section 3.1.3.1). If however the current load imbalance is greater than the specified imbalance value suitable migration candidate is selected. Currently, a simple strategy is used which selects a process of the application on the node with the highest normalized load index. Similarly the destination node is selected with a simple strategie: it is the node with the lowest normalized load index. Finally, CoCheck is requested to migrate the selected process from the source to the destination node.

As the *avenrun* value is influenced by processes which do not belong to the parallel application this approach also can cope with load imbalances caused by external influences. However problems arise when the machine pools of two tuMPI applications overlap. In this case the two load balancers might work against each other. In future versions this situation must be detected and the load balancers must coordinate their migration decisions regarding the nodes in the overlapping node set.

Currently, the above mentioned value for the load imbalance that has to be exceeded so that migrations are actually performed to level the load must be specified by the user as an additional parameter on the command line. As in the case of the measurement intervals (c.f. section 3.1.3.1) this parameter is automatically removed during `MPI_Init`.

3.1.4   Performance Experiments

To evaluate the performance of the automatic load balancer we applied it to a computational fluid dynamics application. This application is briefly described in the next section. After that the hardware environment for the experiments and the results of the experiments are presented.

3.1.4.1   NSFLEX

NSFLEX is a computational fluid dynamics application which is used to solve problems in aerodynamics [5]. More precisely, it solves two and three dimensional Navier-Stokes and Euler Equations. The speed of the flow can stretch from 0.3 to 100 Mach. The discretisation was done with a finite volume method by solving the Reynolds Equations. The linearization uses the Newton method. The turbulent flow is modeled with Baldwin-Lomax and solved with a Gauß-Seidel method. Several grid topologies are supported (C-grid, O-grid and H-grid). The NSFLEX code for MPI which has been used in the experiments was initially implemented for MPICH and solves the Cast-7 problem. The problem was partitioned in such a way, that four processes were required to compute the solution. Each of the processes occupied 6840 kBytes main memory during run-time. Most of this memory (6044 kBytes) was located in the data segment.

3.1.4.2   Hardware Environment for the Experiments

All experiments have been done on five Sun Sparc 10 machines which were equipped with 32 Mbytes of main memory and ran under the SunOS 4.1.3 operating system. The machines were physically distributed over several buildings and were interconnected via the local area Ethernet network of the computer science department. The measurements have been performed during off-peak hours in the nights and on weekends to reduce the influence of other users. Despite these precautions it was not possible to completely dedicate the machines and the network for the experiments.

3.1.4.3 Results

In a first series of experiments we were concerned about how the number of migrations would influence the execution time of NSFLEX. Therefore NSFLEX was executed on four machines and a varying number of migrations were forced to the remaining machine. The average results are depicted in Figure 3.5 whereas $t(n)$ in equation (3.2) is the result of a linear regression applied to all measurements.

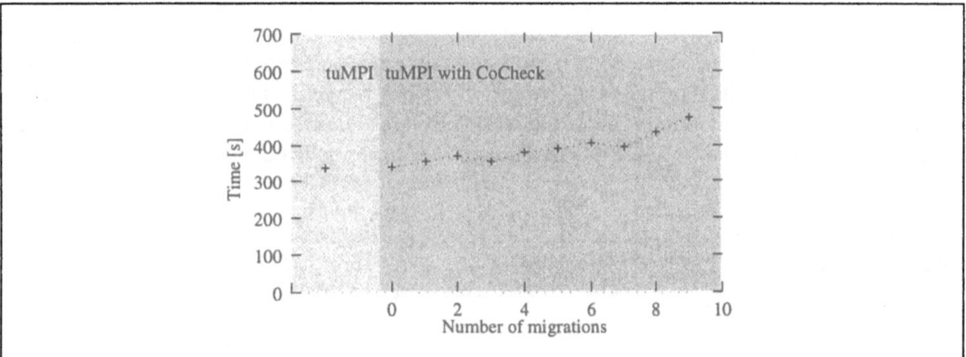

Figure 3.5: Average execution times of NSFLEX depending on the number of migrations.

$$t_{NSFLEX}(n) \quad = \quad 330.55 + 12.83n \tag{3.2}$$

Although the average values and the linear regression show a linear increase of the execution times of NSFLEX a closer look at the individual measurements unveils noteworthy details. In contrast to the expectation that migrations prolong the execution time in any case, surprisingly also reductions could be observed. This is depicted in Figure 3.6 .

The explanation for this effect is as follows. Although the migration itself takes a certain amount of time the experiment already describes an automatic load balancer with a random strategy where migration decisions are based on coincidence. Since the external load could not be completely eliminated, this simple strategy of migrating a randomly selected process lead to better load distributions under certain circumstances. The examination of the log files of the experiments, in which the load values of all machines had been recorded, showed that in the cases where better execution times could be achieved, the source nodes indeed were overloaded.

In a second series of experiments an additional process was placed on one of the nodes where an NSFLEX process was executed. The automatic load balancer was enabled and configured, so that the load imbalance value was set to 1.5 and the two interval parameters were both set to 40 s.

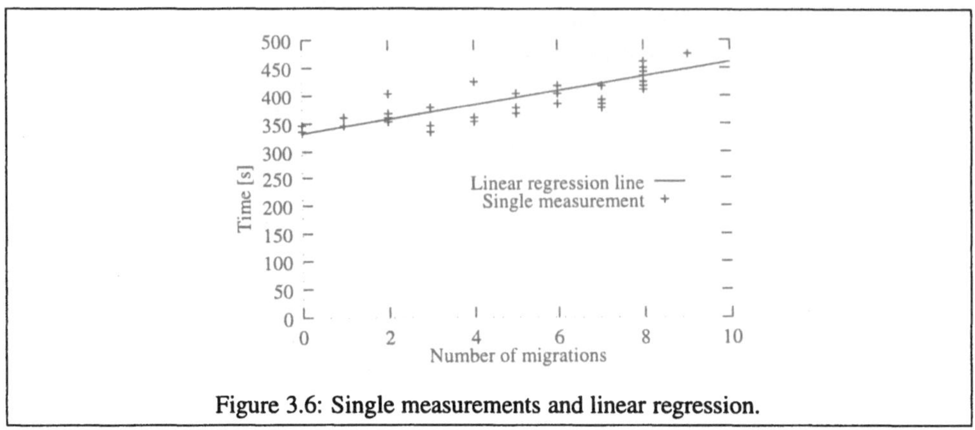

Figure 3.6: Single measurements and linear regression.

Hence, migrations were performed if the normalized difference between the least and most loaded node exceeded 1.5 and the time between two measurements were 40 s independent of migrations having been performed or not. The additional process on the node consumed about the same amount of CPU-time as the NSFLEX process as can be seen from the output of the top command in Figure 3.7.

```
  PID USERNAME PRI NICE SIZE   RES   STATE TIME WCPU    CPU     COMMAND
13402 stellner 76  0    24K    144K  run   1:45 45.37%  45.31% AddLoad
13431 stellner 77  0    6840K  2904K run   1:23 45.37%  45.31% NsFlex
```

Figure 3.7: CPU time of the NSFLEX process and the additional load process.

The average execution times of NSFLEX without performing migrations increased to 697 s in comparison to 336 s without the additional load process. The experiment was repeated with the automatic load balancer being activated. In this case the average execution time could be reduced to 327 s. In some cases more migrations were necessary to achieve an even load distribution. In these cases execution times of 374 s and 500 s respectively could be achieved. The individual measurements and the corresponding average values are given in Figure 3.8 .

In a further experiment with a similar setup as in the previous experiment two additional load processes where placed on two distinct nodes. Here, at least two migrations were necessary to achieve an even load distribution. Although the execution times could not be reduced to the same values from the previous experiment, there were nevertheless performance improvements. A limitation of the current implementation of the decision component is that it only migrates a single process at a given time. Hence, if two processes should be migrated to achieve an even load distribution they are serialized. Therefore, the load remains in an unbalanced condition until the next measurements are performed after 40 s. Compared to the previous experiment the load is 40 s longer unbalanced. The results are summarized in Figure 3.9 .

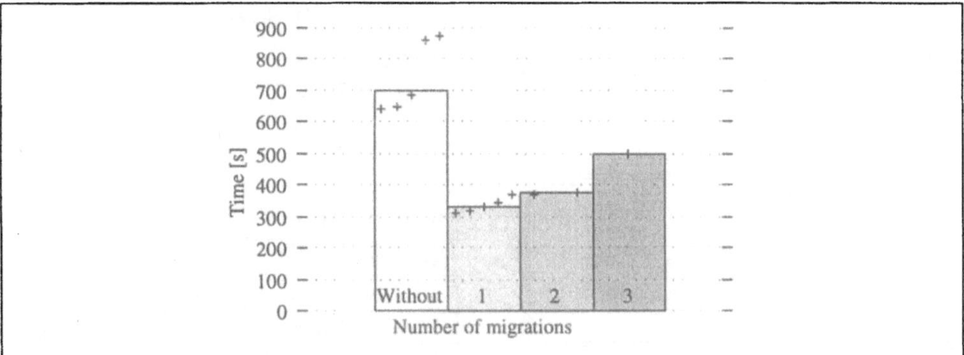

Figure 3.8: Effect of load balancing on NSFLEX execution times with one additional load process.

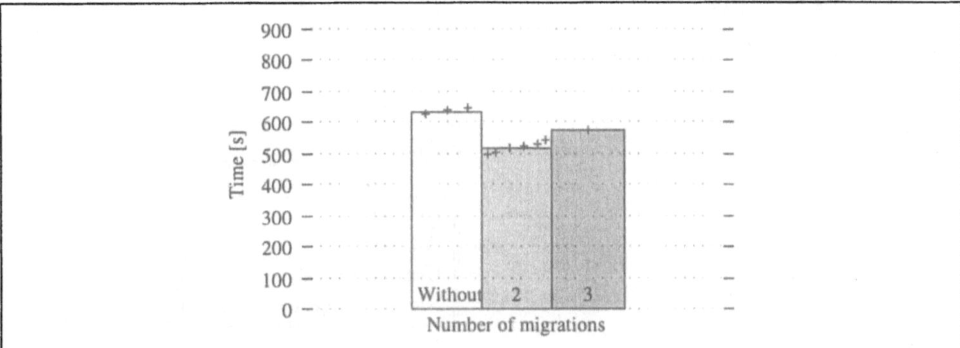

Figure 3.9: Effect of load balancing on NSFLEX execution times with two additional load processes.

3.1.5 Conclusion and Future Work

In concert with tuMPI CoCheck provides the basis for an automatic load distribution system for MPI applications. Both the migration times of processes (not discussed here, refer to [12] for details) and the improvements which could be achieved using the NSFLEX application are very encouraging. Already the simple policy based on the normalized length of the ready queue and the threshold value used in the prototype implementation could reduce execution times from 10% to 54%. A classification of CoCheck according to the terminology given in this book is given in tables 3.1 to 3.2 .

Current limitations of the decision components are that they only initiate the migration of a single process at a given time and that decision components of several applications cannot co-operate. The first limitation reduces the efficiency of distributing the load evenly whereas the latter leads to problems in case of overlapping machine pools when processes migrate to the same machine. Hence, the decision components have to be modified, so that they migrate more than one process if this is appropriate and that they can coordinate their migration decisions in case of

I. General

Objectives	
Domain specific	no
Intent	both
Function	assignment
Integration Level	
Monitoring	runtime system
Strategy	runtime system
Mechanisms	runtime system
Compiler support	none
Structure	
Problem structure	location dependent
Subproblems	computation
Workspaces	processes
Targets	nodes

II. Strategy

System Model	
Model Flavour	physical
Target Topology	NOW, bus, mesh, hypercube, torus, fully connected
Entity Topology	non interacting
Transfer Model	
Transfer Space	systemwide
Transfer Policy	preemptive
Information Exchange	
Information Space	systemwide
Information Scope	complete
Coordination	
Decision Structure	centralized
Decision Mode	autonomous
Participation	partial
Algorithm	
Decision process	dynamic
Initiation	central
Adaptivity	fixed
Cost Sensitivity	none
Stability control	garantueed

Table 3.1: General and Strategy Classification

III. Load Model

Load Index Properties	
Index dimension	single dimensional
Index type	CPU
Aggregation time	comprehensive
Aggregation space	system related
Load Value Composition	
Index combination	none
Weight selection	system characteristics
Model Usage Policies	
Index measurement	periodically fixed
Index propagation	on demand
Model adaptivity	fixed

IV. Migration Mechanisms

Migration Mechanism CoCheck	
Migrant	process
Migration Set Size	limited
Heterogeneity	no
Migration Initiation	immediate
Pre Transfer Media	none
Compression	no
Transfer Policy	complete
Residual Dependencies	none
Migration Transparency	yes
Source Code Transparency	yes

Table 3.2: Load Model and Migration Mechanism of CoCheck.

overlapping machine pools. Furthermore, the intra-application scheduling aspect which is covered by dynamic load distribution should be extended to a resource driven inter-application scheduler which can be found in resource management systems. Finally, the heuristics must be evaluated with more applications. Particularly, the ability to distribute the load in heterogeneous clusters must be examined.

References

[1] P. BRIDGES, N. DOSS, W. GROPP, E. KARRELS, E. LUSK and A. SKJELLUM. *Users' Guide to mpich, a Portable Implementation of MPI*. Argonne National Laboratory, 9700 South Cass Avenue, Argonne, IL 60439-4801, July 1995.

[2] D. L. EAGER, E. D. LAZOWSKA and J. ZAHORJAN. Adaptive Load Sharing in Homogeneous Distributed Systems. *IEEE Trans. Softw. Eng.*, SE-12(5):662–675, May 1986.

[3] A. GEIST, A. BEGUELIN, J. DONGARRA, W. JIANG, R. MANCHEK and V. SUNDERAM. *PVM: Parallel Virtual Machine — A Users' Guide and Tutorial for Networked Parallel Computing*. Scientific and Engineering Computation. The MIT Press, Cambridge, MA, 1994.

[4] Genias Software GmbH, Erzgebirgstr. 2B, D-93073 Neutraubling, Germany. *CODINE User's Guide*, 1993.

[5] M. LENKE, S. RATHMAYER, A. BODE, T. MICHL and S. WAGNER. *Parallelization with a real-world CFD application on different parallel architectures*, volume 2, chapter 8, pages 119–166. Computational Mechanics Publications, Southampton, Boston, 1995.

[6] M. LITZKOW, M. LIVNY and M. MUTKA. Condor — A Hunter of Idle Workstations. In *8th Int. Conf. on Distributed Systems*, pages 104–111, Los Alamitos, CA, 1988. IEEE Computer Society Press.

[7] J. MENDEN and G. STELLNER. Proving Properties of PVM Applications — A Case Study with CoCheck. In A. BODE, J. DONGARRA, T. LUDWIG and V. SUNDERAM, editors, *Parallel Virtual Machine — EuroPVM'96*, volume 1156 of *Lecture Notes in Computer Science*, pages 134–141, Berlin, October 1996. Springer Verlag.

[8] MESSAGE PASSING INTERFACE FORUM. MPI: A Message Passing Interface Standard, May 1994.

[9] G. STELLNER, A. BODE, S. LAMBERTS and T. LUDWIG. NXLib — A Parallel Programming Environment for Workstation Clusters. In C. HALATSIS, D. MARITSAS, G. PHILOKYPROU and S. THEODORIDIS, editors, *PARLE'94 Parallel Architectures and Languages Europe*, volume 817 of *Lecture Notes in Computer Science*, pages 745–748, Berlin, July 1994. Springer Verlag.

[10] G. STELLNER and J. PRUYNE. Resource Management and Checkpointing for PVM. In *2nd European PVM Users' Group Meeting*, pages 131–136, Lyon, September 1995. Editions Hermes.

[11] G. STELLNER. Consistent Checkpoints of PVM Applications. In *1st European PVM Users Group Meeting*, http://www.labri.u-bordeaux.fr/~desprez/CONFS/PAPERS/abs010.ps.gz, 1994.

[12] G. STELLNER. CoCheck: Checkpointing and Process Migration for MPI. In *Int. Parallel Processing Symposium*, pages 526–531, Honolulu, HI, April 1996. IEEE Computer Society Press, 10662 Los Vaqueros Circle, P.O. Box 3014, Los Alamitos, CA 90720-1264.

3.2 Load Distribution Strategies of the Distributed Thread Kernel DTK

by Niels Reimer and Markus Pizka

3.2.1 Introduction

The Distributed Thread Kernel DTK evolved as part of the project MoDiS (Model-oriented Distributed Systems). MoDiS is best characterized as a top-down driven, language and object-based approach to construct an holistic distributed system, integrating applications and operating system (OS) services [1]. Starting from abstract concepts and models for the construction of distributed systems we developed the distributed programming language INSEL (INtegration– and

Separation–supporting Language), which is an imperative object-based and type-safe high-level language with an Ada–like syntax [2]. It supports the specification of explicit parallelism by providing concepts for active objects (*actors*) in addition to passive ones. The programmer specifies parallel computations on a high level of abstraction by creating actors as instances of self-defined actor classes. Hence, INSEL allows the determination of parallelism without any references to the OS or the physical execution environment such as number of processors or physical nodes.

On the one hand, this transparency enables adaptability of INSEL applications to varying configurations and states of the hardware as well as it eases the development of distributed applications. On the other hand, it requires a flexible and well adapted underlying resource management system to achieve efficient execution. This comprises compilation techniques, the runtime system and basic OS services.

With two prototype implementations of a runtime system for INSEL we were able to gather experiences on how to match the demands of active and passive INSEL objects to underlying host OSs. Our prototype EVA (Experimentalsystem für Verteilte Anwendungen) was implemented on a cluster of HP PA-RISC workstations running HP-UX to investigate realisation techniques for INSEL actors. The requirements of these active objects to the resource management layer are characterized by varying granularity[4], dynamic creation and deletion and a dynamically changing number of active objects which are simultanousely executing.

Due to the somehow *heavy–weight* nature of UNIX processes, they can usually not serve as an adequate realization technique for such language potentialities. Therefore, we developed the distributed thread kernel (DTK) [3] to bridge the gap between OS features and the demands of INSEL applications. We decided to develop it as a user–level threadpackage[5] with integrated distribution strategies, that assign threads by an initial placement to the individual nodes of a cluster of workstations.

We performed extensive benchmarks considering the aspects of load distribution and the impact of different strategies. After presenting the basic structure of a DTK–system we will focus on DTK's strategies and their load related aspects. We will briefly describe DTK's different strategies and classify them according to the classification schemes of chapter 2.1 on page 13. These strategies will be compared in a summarizing section before we conclude with a short description of our future work. The results of our measurements are presented in [3] for the interested reader.

3.2.2 DTK Basics

DTK is a layer on top of the local UNIX–OS of each workstation. It is implemented as a C++ library which has to be linked to the executable of the distributed application. On each node there is exactly one DTK–task running, which is a UNIX-process comprising a local DTK (DTK–lib) and the local parts of the distributed application. All DTK–tasks together build the distributed runtime environment that provides a transparent and virtually uniform computing resource to the user. Figure 3.10 illustrates the general structure of a DTK–system running on top of a UNIX–OS.

DTK offers five strategies to select a DTK–task for the creation and execution of new threads – *Random, OptimalLocal, Load, RBidding* and *SBidding*. Mapping strategies are DTK's only means

---

[4]complexity concerning both, memory and calculation aspects
[5]as a light–weight alternative to UNIX–processes

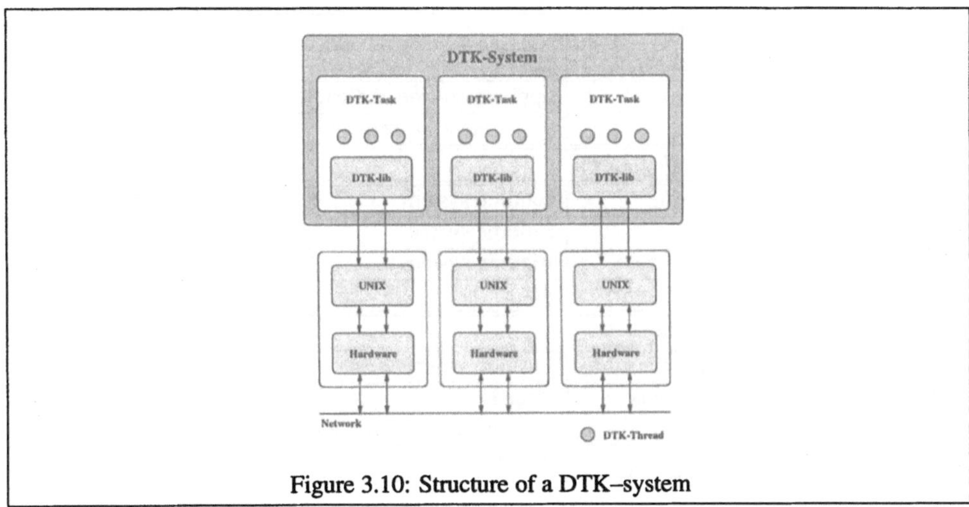

Figure 3.10: Structure of a DTK–system

to achieve an even load distribution. Since we aim towards finegrained problem decomposition we expect active INSEL objects to perform mostly small computations in parallel. Therefore the lack of support for thread migration does not seem to be a severe limitation in the INSEL context.

Individual performance values are determined for each node to reflect the non–uniform performance of our target hardware. These values are assumed to be static and are forwarded to the DTK–tasks at setup–time of the distributed runtime system.

3.2.3 General Classification

Since the general classification scheme acts on a high level of abstraction the classification is the same for all presented strategies. Additionally, this scheme covers only system–related aspects leaving the strategy details to the strategy classification. Here we present the classification as a table accompanied by some additional explanations if needed.

The items of the objectives subsection should be self–explanatory. Since we regard the DTK and its integrated thread mapping strategies as part of a *runtime system* the related integration level classification is easy. However the structural aspects should be explained in more detail. The load distribution refers only to the active components which are realised as threads. Therefore the terms application objects and threads are both given to characterise the workspace in table 3.3 .

The following subsections introduce and explain the different strategies. The bidding strategies are much more complex than the strategies Random, OptimalLocal or Load. The basic idea of bidding algorithms is described in detail in [5]. With the work of [3] a modification of the basic idea is implemented. The classification of all strategies and their individual load index aspects are given afterwards. In general the load index is the number of threads in the run–queue.

Objectives	
domain specific	no
intent	application-oriented exclusive
function	assignment, placement
Integration Level	
monitoring	runtime system
strategy	runtime system
mechanisms	runtime system
compiler support	none
Structure	
problem structure	location independent
subproblems	threads
workspaces	threads / application objects
targets	nodes

Table 3.3: General Classification Scheme for DTK and its strategies

3.2.4 Random

Every newly created thread is mapped to a randomly chosen DTK–task.The decision process is simply picking up a random number out of an interval. This interval consists of adjacent sub–intervals, one for each available DTK–task. The sub–interval which contains the random number is then used to determine the target DTK–task. To consider fairness aspects the sizes of these sub–intervals are proportional to the performance values of the nodes (and their DTK–tasks, respectively).

This simple strategy does not use any runtime information (load index, application behaviour, ...) and is often taken as a reference strategy. In the following the strategy random is abbreviated by RA.

3.2.5 OptimalLocal

The OptimalLocal strategy (OL) considers runtime information. The algorithm works in the following way: every DTK–task keeps track of the running threads spawned off from the local DTK–task onto remote DTK–tasks. With the assumption that all spawned threads create the same workload the strategy searches for the minimum quotient of number of children (threads) and the performance value of that DTK–task. The idea is to burden the entire system with a mostly even (optimal) distribution of jobs (running threads) considering only the local point of view. The view is restricted because the mutually created workload of two other DTK–tasks will not reflect in the local data structure.

At some points this strategy is difficult to fit into the given classification scheme. We label the decision mode as *autonomous* because the designated target node can not refuse to accept the additional work and the decision is made without special negotiations. A real load index is not obtained by measurement but derived approximately from the number of spawned child threads onto the remote DTK–tasks. This method implies sending an acknowledge message back to the initiating DTK–task on termination of a child.

3.2.6 Load

The actual local load index (number of running threads) of each DTK–task is periodically updated. If the index has changed beyond certain thresholds an appropriate message is sent to all other DTK–tasks (via a multi–cast) to update the load value in their data structure. Every time a new thread has to be created the DTK–task with the minimal quotient of load value and performance value is chosen. After sending the order to create and execute a thread to a remote DTK–task the local entry of the load value of the target DTK–task is incremented.

Despite of the overall cooperative behaviour we label the strategy to be *autonomous* for the same reasons as mentioned above. Since the meaning of the updating message can be interpreted as an overload signal as well as a request we classify it as *sender–receiver* (abbreviated as send–rec.) initiated. A stability control is *not required* since the strategy is only invoked by time–outs or at thread creations. Undoubtedly it is possible to decrease the time–out interval to a rate that shifts the major action to the repeated load index determination. But even then the amount of messages sent will be low, since the new load index is only broadcasted if it has significantly changed. Therefore an explicit stability control of the load distribution strategy is *not required.*

3.2.7 RBidding

The idea of receiver initiated bidding strategies (like RBidding (RB) is one) is that nodes with a surplus of capacity (so–called idle nodes) broadcast requests for work. A DTK–task is regarded as idle if the local load index drops below a threshold and is considered as overloaded if the local load index is above the same threshold. In general, overloaded nodes react to this offer by passing a fraction of their work to the requesting nodes, thus leading to migration. Since we do not support migration, our overloaded DTK–tasks charge requesting idle DTK–tasks with the creation and execution of new threads. If there is no request for work available the new thread creations are delayed and their requests are stored in a queue. To follow the principles of the receiver initiated bidding each DTK–task has to maintain a timer, a queue of threads to be created and an array of flags representing the requests for work of the remote DTK–tasks. Besides the current workload index the value of the most–recent load index is also stored to observe changes of the local load. The following events result in activities of the load distribution strategy:

- Thread creation
- Time–out
- Request for work
- Request for work invalidation message

The actions related with each event are now described in more detail.

Thread creation: When a thread has to be created the load index is determined and if it is below a threshold the thread is created locally and the load index is increased. In case the load index is above the threshold the strategy searches through the array of request flags, invalidates the first request, increases the locally stored index of the requester and finally sends the new thread creation order to the requester. If there are no requests for work the strategy enqueues the thread creation request onto the local queue.

Time–out: Some action is also taken when a time–out is encountered. First the load index is determined. If the current load index and the most–recent load index is above a certain threshold the timer is reset and no further actions are started.

If the current index is above the threshold and the most–recent index was below it then a multi-cast message is sent to other DTK–tasks to invalidate old requests for work of this DTK–task.

If the current load index is below the threshold this DTK–task will first try to satisfy local pending threads and then ask other DTK–tasks for work. This is implemented in the following way: if the most–recent load index is below the threshold old requests for work have to be invali-dated first by sending a message to all DTK–tasks. Then the local DTK–task processes the queue of threads waiting for their creation.

With each creation the current load index is incremented. This happens until the index reaches the threshold or the queue becomes empty. The action for the first case is then terminated by resetting the timer. In the second case the load index is incremented and a request for work is sent to the next DTK–task according to a common cyclic order imposed on all DTK–tasks of the system. This action is repeated until the work load index has reached the threshold again. Finally the timer is reset.

Request for work: If a DTK–task receives a request for work and the local queue is empty then the appropriate flag is set. If the queue is not empty then one request for a thread creation is dequeued and sent to the requesting DTK–task.

Request for work invalidation message: If a DTK–task receives an invalidation message the appropriate flag in the request for work array is cleared.

The RBidding strategy shows some special properties. The transfer space is *global* but in general the information scope is *restricted* (not all DTK–tasks will receive a request for work when a resource is idle). Even if we feel the *microeconomic* model flavour of the strategy it is not competitive since the receivers of a request for work do not have to compete. The requester only offers the amount of capacity that he can satisfy at the moment assuming that no additional threads were created locally which would reduce the local capacity. Hence, RB only provides *low* stability control. To be precise the load index is never propagated among the DTK–tasks but the arrival of requests and invalidation messages can be regarded as an expression of remote load situations. The behaviour of a request (searching for work *on demand* or to be *adaptive* to a sudden idleness) justifies our classification.

3.2.8 SBidding

The principle of a sender initiated bidding strategy (like SBidding (SB)) is that overloaded nodes broadcast a request to share their load with idle nodes. Idle nodes reply and get some jobs passed. In general this method assumes the use of migration techniques. In our case overloaded DTK–tasks delay the thread creation, broadcast requests for idle DTK–tasks and pass the thread creation and execution on reply.

The sender initiated load distribution strategy SB also keeps a queue for the threads to be created and a timer. Actions are taken on occurrence of the following events:

- Thread creation
- Time–out
- Request for idle DTK–tasks
- Acknowledge message from an idle DTK–task

The related actions with each event are now described in more detail.

Thread creation: On thread creation the load index is determined and if it is below a threshold the thread creation is performed locally. If the index is above the threshold the thread creation is delayed and therefore enqueued. If the queue was empty before, a request for an idle DTK–task is sent to the other DTK–tasks as a multi–cast message. If there were already threads pending no further action is performed.

Time–out: If a time–out is encountered the load index is updated. If it is above a threshold and the queue contains waiting threads then a request for idle DTK–tasks is sent and the timer is reset. If the queue is empty only the timer is reset. In case the load index is below the threshold then the entries in the queue are processed locally. With each thread creation the load index is incremented. This is repeated until the queue is empty which finally causes a timer–reset or until the load index is above the threshold which branches to the actions described first in this paragraph.

Request for idle DTK–tasks: On an incoming request for idle DTK–tasks the current load index is examined. If the index is above the threshold the request is ignored. If the index is below the threshold the load index is incremented and an acknowledge message is sent offering capacity to the requester. The increase of the load index is done beforehand to counteract situations when a sudden flood of requests for idle resources occurs. This enhances the stability of the strategy.

Acknowledge message from an idle DTK–task: If the request for idle resources is matched by an acknowledge message the queue is checked. If it is empty the message is ignored, otherwise a thread is dequeued, the appropriate load index is incremented and the remote creation of the thread is performed.

The remarkable property of the SBidding strategy compared with the RBidding is, that it is a *competitive* strategy, since the load distribution on an acknowledge and the emitting of an acknowledge is done according to a first–come–first–serve policy.

3.2.9 Comparison of DTK's strategies

A summary of the strategies is given by their classifications in table 3.4 . The classifications of their load indices are presented in table 3.5 on page 89. The strategy RA is omitted in this table because it lacks a load index.

The interested reader may refer to [3] for a detailed presentation of benchmarks with different thread creation patterns. Here we will focus on the major results and present the derived conclusions. All strategies with a load index show better results than the reference strategy RA. For fine–grained parallel applications OL and the two bidding strategies lead to a better performance of the system, but the tunable parameters (e.g. time–out value) require reasonable consideration and adjustment. One reason for the success of RB stems from the fact that only this strategy managed it to create the majority of threads locally. Further improvements are obtainable with the use of strategies that are application–oriented because these strategies would consider the existance of communication among the objects and take the resulting communication costs into account.

Strategy	RA	OL	LO	RB	SB
System Model					
Model Flavour	random	none	fairness	micro-economic	micro-economic
Target Topology	het. NOW	het. NOW	het. NOW	het. NOW	het. NOW
Entity Topology	indep. tasks	indep. tasks	indep. tasks	indep. tasks	indep. tasks
Transfer Model					
Transfer Space	systemwide	systemwide	systemwide	systemwide	systemwide
Transfer Policy	non-preem.	non-preem.	non-preem.	non-preem.	non-preem.
Information Exchange					
Information Space	none	systemwide	systemwide	systemwide	systemwide
Information Scope	none	partial	complete	partial	complete
Coordination					
Decision Structure	distributed	distributed	distributed	distributed	distributed
Decision Mode	autonomous	autonomous	autonomous	cooperative	competitive
Participation	partial	partial	partial	partial	partial
Algorithm					
Decision process	static	dynamic	dynamic	dynamic	dynamic
Initiation	sender	sender	send–rec., timer-based	send–rec., timer-based	sender, timer-based
Adaptivity	fixed	adjustable	adjustable	adjustable	adjustable
Cost Sensitivity	none	none	none	low	low
Stability control	none	none	not required	partial	partial

Table 3.4: Strategy classification of Random, OptimalLocal, Load, RBidding and SBidding

Therefore, creation of threads should sometimes be forced to be performed locally even if the node is well loaded.

3.2.10   Related Work

Besides EVA/DTK we implemented another prototype of an INSEL runtime–system, called AdaM. It runs on a network of i486/Mach3 workstations on top of the Mach3.0 micro–kernel. The main objective of this project was to investigate distributed memory management.

Similar to load management, we found that there is no uniform optimal way to realize access to passive objects in a distributed environment. Hence, to allow adaption to different situations and system states, three alternatives are provided by the runtime–system (completely transparent for the INSEL programmer) — object replication, object migration and access via RPC. A heuristic is used to choose one of these techniques dynamically at runtime. Object access patterns and frequencies as well as the costs for replication are used as input for the heuristic. Observing these characteristics we obtain an application–oriented realisation of access to shared data objects which has a great influence on the performance [6].

Strategy	OL	LO	RB	SB
Load Index Properties				
index dimension	single-dim.	single-dim.	single-dim.	single-dim.
index type	CPU	CPU	CPU	CPU
aggregation time	instantaneous	instantaneous	instantaneous	instantaneous
aggregation space	system related	system related	system related	system related
Load Value Composition				
index combination	none	none	none	none
weight selection	not applicable	not applicable	not applicable	not applicable
Model Usage Policies				
index measurement	on demand	on demand period. fixed	on demand period. fixed	on demand period. fixed
index propagation	none	adaptive	on demand adaptive	on demand adaptive
model adaptivity	fixed	fixed	fixed	fixed

Table 3.5: Load Index Classification of OptimalLocal, Load, RBidding and SBidding

3.2.11 Conclusion & Future Work

One of the lessons learned from the research done with DTK is that distributed load management is essential and achievable even with simple and intuitive strategies. Nevertheless the impact on the system performance using application–oriented resource management will be important as investigated in AdaM. A strong orientation and adaption towards the applications' demands will further improve the benefits of load management. In addition to that, it should be obvious, that because of the strong influences of load distribution on memory management and vice versa, both tasks should not be viewed isolated. Hence, in future works we have to combine load and memory techniques to an integrated resource management.

In our current work we are developing an integrated INSEL system, comprising applications as well as classical OS services. In contrast to our prototype implementations this system will not run as a runtime–system on top of a host OS (UNIX, e.g.) but be based on an own micro–kernel. At the same time, we are implementing tools, such as a compiler and a dynamic linker and loader that allow an extensibel and incremental construction of the system. All of our tools, the services of the OS and the micro–kernel are being specifically tailored to the demands of our language concepts.

From DTK and AdaM we learned, that different alternatives for the accomplishment of load and memory management tasks have to be provided to allow adaptable and efficient resource management. Decisions between different realization techniques have to be based on application–specific knowledge. To meet this requirement, static application–specific properties will be analyzed by the INSEL compiler by exploring structural dependencies [4]. Dynamic properties, such as communication frequencies will be monitored during execution. The resource management layer of our system is designed to utilize these static and dynamic application–level information to choose an adequate realization technique matching the application–object's requirements. Hence, resource management will be application–oriented with information derived from analyses at compiletime and runtime.

References

[1] M. PIZKA. A Language-Based Approach to Construct Structured and Efficient Object-Based Distributed Systems. In *30th Hawaii Int. Conf. on System Sciences.* IEEE, 1 1997.

[2] R. RADERMACHER and F. WEIMER. INSEL Syntax–Bericht. Technical Report, TU Munich, 1996. TUM–I9617, SFB–Bericht Nr. 342/08/96 A.

[3] R. RADERMACHER. *Eine Ausführungsumgebung mit integrierter Lastverteilung für verteilte und parallele Systeme.* Ph.d. thesis, TU Munich, 1995.

[4] P. SPIES, C. ECKERT, M. LANGE, D. MAREK, R. RADERMACHER, F. WEIMER and H.-M. WINDISCH. Sprachkonzepte zur Konstruktion verteiler Systeme. Technical Report, TU Munich, 1996. TUM–I9618, SFB–Bericht Nr. 342/09/96 A.

[5] J. STANKOVIC and I. SIDHU. An Adaptive Bidding Algorithm for Processes, Clusters and Distributed Groups. In *4th Int. Conf. on Distributed Systems*, pages 49–59, 1984.

[6] H.-M. WINDISCH. *Speicherverwaltung für konzeptionell strukturierte verteilte Systeme.* Ph.d. thesis, TU Munich, 1996.

3.3 *Dynasty*: Economic-Based Dynamic Load Distribution In Large Workstation Networks

by Martin Backschat and Alexander Pfaffinger

3.3.1 Introduction

In the field of parallel high performance computing, there is a trend towards workstation clusters, because they are available at many sites. On the one hand their computing capacity is huge but the workstations are commonly used as desktop computers for text processing or compilations and are usually idle overnight, which barely exploits their available potential. On the other hand a network of workstations is comparatively cheaper than a dedicated parallel computer offering therefore a cost-effective alternative for parallel computing.

Nowadays, many clusters with many different powerful workstations form a large heterogeneous distributed computer system. The system's topological and architectural variety and the unpredictable dynamic runtime behaviour of the parallel applications make their manual or static distribution insufficient. To overcome these problems dynamic load distribution is used.

Load distribution systems try to keep all computing nodes busy by migrating tasks from overloaded to underloaded nodes. If the load of a node drops below a certain threshold the node becomes underloaded. Another threshold determines when a node is considered overloaded. If underloaded nodes cannot be supplied with new tasks fast enough they become idle. To avoid this situation, *load balancing systems* try to balance the number of tasks on all nodes. This usually results in more migrations and a more frequent exchange of load information [2; 11], which are major drawbacks in large computer networks due to the resulting overhead.

A key issue in the design of such systems is whether the responsibility of task distribution resides in one dedicated node or is physically distributed among all machines. In *decentralized* systems each node exchanges load information and migrates tasks only to its neighbours or at least to a small number of nodes, which makes this approach scalable. But there are also several disadvantages, e.g. optimal global distribution is impossible due to the local view. Since tasks are only migrated in a restricted scope, decentralized load distribution systems cannot keep pace with "task explosions", typically occuring in start-up phases. Furthermore, the decentralized control may cause instabilities and migration oszillations [7; 15].

These problems are overcome by *centralized* systems. One dedicated node keeps track of the network topology, collects all load information and initiates migrations. But usually, this form of organization is sensitive to failures of the central node and of course not scalable.

Hierarchical organized systems, like *Dynasty*[1], combine the advantages of both approaches. They are *scalable* and therefore especially appropriate for large distributed computer systems [16]. Hierarchical organization is already used successfully in related areas such as Internet routing via OSPF [RFC1247] and Internet domain name service [RFC1032].

In recent years much effort has been spent in developing new load distribution mechanisms that are inspired by analogies from other areas, e.g. fluid mechanics [8] and especially economics [3; 5; 9; 17]. Since market devices such as auctions and prices facilitate resource management in human societies, one might expect them to be similarly useful in distributed computer systems — a proposal which has been elaborated in several projects such as *Spawn* [18] and *Enterprise* [13]. Considering this analogy, tasks are equipped with funds to finance their migrations and executions. The price mechanism allows machines with different capabilities to have different values, enabling tasks to flexibly devote their currency to the resources most important to them. In *Spawn* and *Enterprise* machines hold auctions in order to sell processing time to competing tasks. Bids are only accepted from local and adjacent tasks. This results in the same drawbacks as in traditional decentralized systems.

In *Dynasty* we take a similar economical motivated approach, but instead of auctions we use brokering of machines requested by tasks and carried out by a hierarchical organized broker system. The prices for allocating computing resources are dynamically fixed, based on measured performance and system load, and avoid overloading of interactively used computers (*non-intrusiveness*). In contrast to the *Spawn* system we also consider data exchange and migrations as services that tasks have to pay for.

Another key issue in designing load distribution systems is *transparency*. Some systems are implemented as part or extension of an operating system [12; 14; 20], working fully transparent to the applications and their users. This approach assumes that each participating node runs the same operating system or extension. Other systems consist of libraries linked to the parallel application or/and of separately working components (daemons) [6; 20]. They provide the application with recommendations of where to place or migrate tasks and offer transport mechanisms for migrating tasks and data. Therefore, some load distribution logic has to be coded into the application, which is usually a problem with commercial products or FORTRAN code. On the other hand this approach can take full advantage of the knowledge about the application and its runtime behaviour.

Our system uses separately working components, one per computing node, called local brokers. In order to achieve transparency at the application level and avoid code modifications, all adaptions are hidden in the application server, a module that is linked to the application, providing an interface to *Dynasty*. This approach allows many different applications to use *Dynasty* for load distribution (*reusability*), even simultaneously.

In *Dynasty* we focus on divide&conquer applications that run the same program on each computing node. The nodes simultaneously work on different tasks. This computing model is known as SPMD (*single program multiple data*).

3.3.1.1 Contributions

The concept and design of our dynamic load distribution system contributes to several areas of research. The key contributions are summarized below:

- First implementation of a broker-based load distribution system working on economical principles.
- Application of distributed artificial intelligence agents: The brokers are reactive, autonomous, cooperative and have distributed global knowledge.
- The use of economical and monetary strategies in application-level load distribution and task splitting: Monetary funding and economical reasoning controls the runtime behaviour of the application.
- Application of dynamically priced (distributed) services based on offer/supply and demand for resource allocation.

In the next section, we give a comprehensive overview over the *Dynasty* load distribution system, where we introduce the main components and concepts of the project. In section 3.3.3 on page 98 we present the results of a performance evaluation of the current implementation of *Dynasty*. We chose a parallel Finite Element solver that is distributed on up to 96 workstations. And finally in section 3.3.4 on page 100 we present a complete classification of *Dynasty* according to the schemes of Chapter 2.

3.3.2 Concepts and Design of *Dynasty*

Dynasty implements a computational economy. The costs for services offered by the *brokers* and for using *computing resources*, e.g. workstations or nodes of a multi-processor system, are based on supply and demand. *Tasks* demand these services and utilize computing resources. The following subsections will present these three economical entities and their relation in more detail.

3.3.2.1 Overview

In order to pay its expenditures each task has its own funds. Expenditures include: rent for utilizing computing resources, brokerage for getting assigned to a target host and fees for migration and data transport services. The root task's funds are updated on a regular basis. It supplies all of its children tasks with income. The children tasks, in turn, forward part of their income to the next lower level of the task tree.

After having paid rent in advance a task is scheduled to the application program on that machine. While processing a task the application server may split off new children tasks and provide them with their own monetary funds. Furthermore, depending on the current task's economical situation, it may contact the local broker requesting the assignment of a new host for the child's subsequent execution.

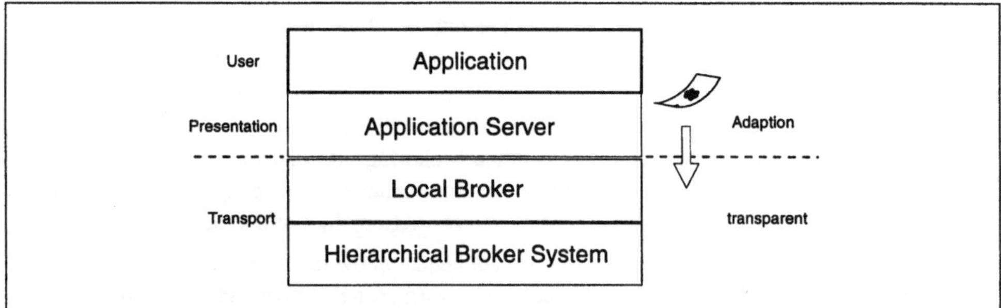

Figure 3.11: The *Dynasty* system consists of four layers, including the application and its frontend, the application server.

The rent is periodically fixed. It is based on several criteria such as the task's computational complexity, the host's capabilities, local load, the current average global performance and rent. It is the basic instrument for regulating a machine's load.

The assignment of tasks to machines (brokering) is carried out by the hierarchical organized broker system, the core of *Dynasty*. The virtual brokers act similar to real estate brokers. They accumulate rent-, performance-, and capability-information about machines that are under their administration. If an assignment is demanded, the broker system selects a target host that conforms to the task's resource and performance requirements and its income. This is accomplished by reference to all available information. The broker system then fixes the brokerage to be paid by the migrating task.

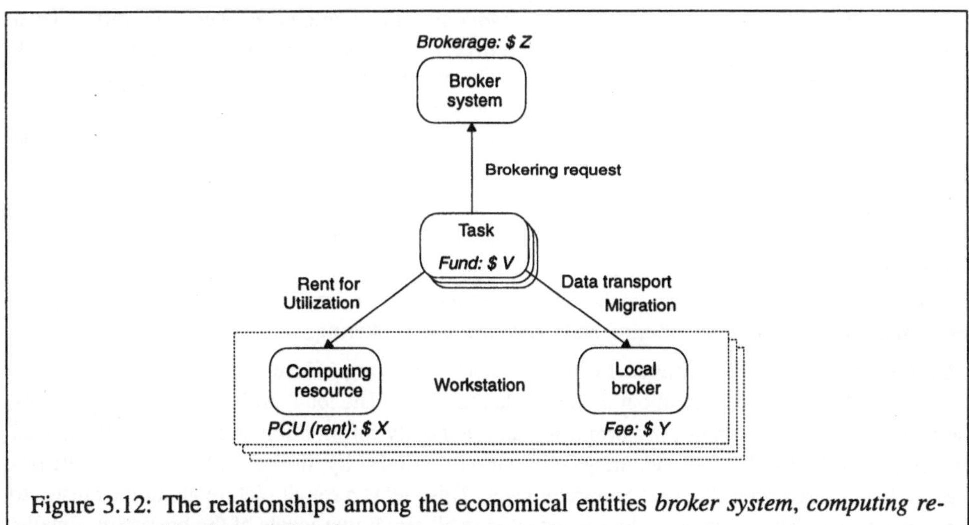

Figure 3.12: The relationships among the economical entities *broker system*, *computing resources*, *local brokers*, and *tasks*.

The more money a task is willing to pay for its assignment the more levels of the broker hierarchy take part and the larger is the number of machines under consideration (*selection range*), ranging from cluster-wide (one level) up to system-wide (all levels). The task is migrated as soon as a target host is selected. This is carried out transparently by the hosts' local brokers and again has to be paid for by the task.

Because the brokers are independent from specific applications they can serve requests of different applications at any time. Our broker session protocol defines the interaction of the brokers, i.e. the brokers' exterior behaviour in cooperative brokering and distributed information storage. Therefore, owners of workstations or network administrators may design their own brokers, which, for example, consider local particularities and restrictions like online periods.

The efficiency of the dynamic load distribution depends on the tasks' ability to take advantage of their funds. Funding may be used, for example, to reward cost-effective and productive children tasks by raising their income. The amount of a task's monetary income may also depend on its priority or computational complexity. A task may initiate its migration if its income is less than the local rent, its funds exceed a certain limit, or nodes with better performance and appropriate rent are available. All these decisions are part of the economical reasoning that has to be coded into the application server. The application server is the adaption layer between the economical system and the unmodified application code. It is provided as a source code skeleton to be filled out by the application programmer.

3.3.2.2 The Broker system

The brokers are responsible for assigning tasks to machines. To implement this service efficiently, we have chosen a well suited form of broker organization, and developed appropriate algorithms for information exchange and brokering.

- **The broker hierarchy:** The brokers are organized as a cooperative, hierarchical broker system. The number of levels depend on the specific network size and topology. Brokers on a certain level exchange data only with their father on the next higher level and their sons on the previous level, thus keeping communication locally restricted. The hierarchical system has many advantages. It is, for example, easy to integrate a new subnetwork, even at runtime. One simply has to register the new network's top broker to a broker of an appropriate level in the hierarchy. It is also possible to increase or even decrease the number of levels of the broker hierarchy at runtime.

The local brokers are the lowest level of the hierarchy. One local broker is running on each machine. They schedule tasks to application servers, fix local rent, and interact with their cluster brokers. Furthermore, they offer several services to local application servers, as is described in the next section. The cluster brokers make up the first level of the inner broker system, i.e. they are responsible for information gathering and exchange, and brokering.

- **Broker information exchange:** In order to keep brokering efficient the broker system uses regularly updated information about all available machines. The local brokers periodically supply their cluster brokers with rent-, performance-, and capability-information about their workstations. The cluster brokers collect all incoming information. Periodically, they calculate several statistics that are then passed on to the next hierarchy level. This yields a rough picture about the cluster to the upper broker. The same procedure is repeated in all remaining levels of the hierarchy. Finally, global information reaches the root broker, who, in turn, sends it downwards to all brokers in the next lower level on a regular basis. Therefore, all brokers gain local and current global knowledge. The global statistics are accurate and sufficiently up-to-date to ensure reliable brokering.

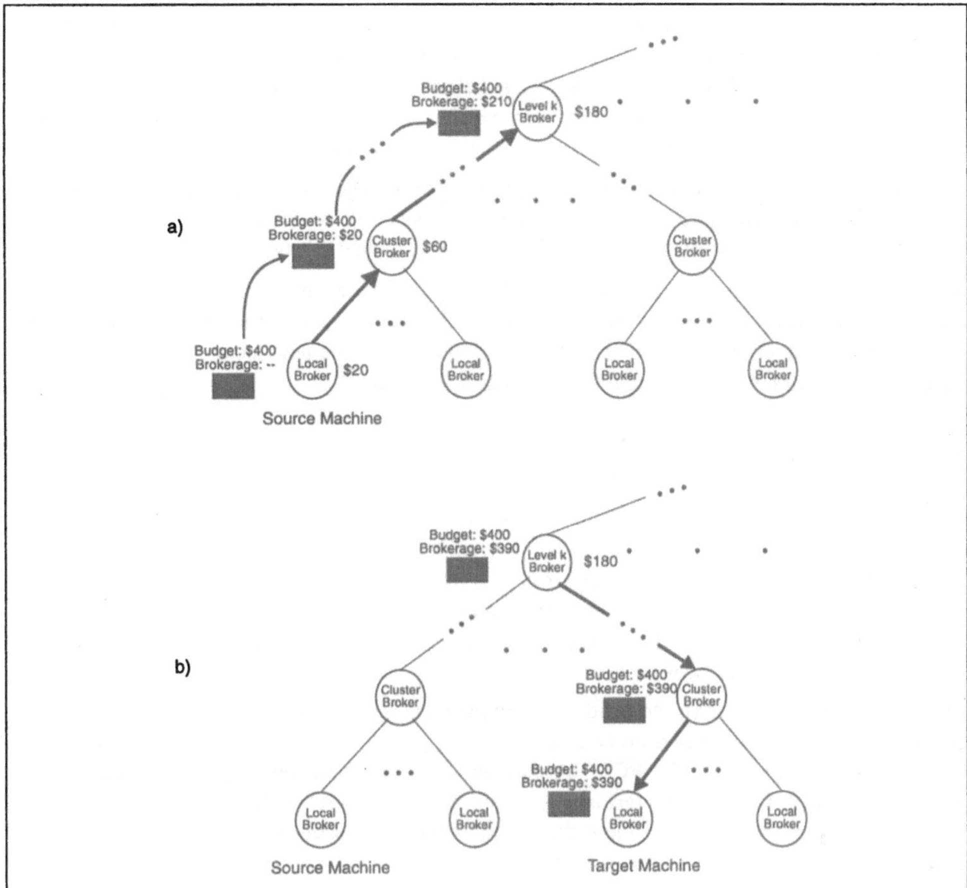

Figure 3.13: Brokering: a) The request is forwarded as long as the accumulated brokerage does not exceed the budget, here: up to level k. b) In each level the current broker selects one of its subbrokers. Finally, the target host is selected.

- **Brokering:** Requests for brokering are issued by application servers and sent to their local broker. The request is forwarded upwards in the hierarchy, each broker on the request's path demanding a certain brokerage (see Figure 3.13).

A request contains the amount of money the task is willing to pay for brokering, the task's computational effort, and a list of requirements concerning the desired target host, e.g. the minimum performance and memory capacity. Additionally, the request provides information about the task's funds and income. By declaring a computing preference a task specifies the relative importance of computing performance versus rent.

The brokerage depends on the broker's level and on its throughput of requests. If the accumulated brokerage exceeds the request's budget the current broker begins with the procedure of brokering. Therefore, it evaluates the qualifation of each of its subbrokers, considering the task's requirements and preferences. The better the qualification the more probable the subbroker will be selected. The probabilistic selection takes into account information age and the number of suitable workstations administered by the subbroker. The selected broker repeats this procedure with its own subbrokers and so on. Finally, the request reaches a local broker, the selected target, who sends an acknowledgement to the task's current local broker immediately.

Usually the application server fixes the maximum brokerage value depending on the task's computational effort and importance. Very important tasks will be granted system-wide choice while simple computations will be kept locally or within the local cluster.

• **Migration:** After having received the target host's acknowledgement the task migrates. The local broker is responsible for transfering the task to the target host and for redirecting future input data of the task's father and result data of the task's children to the new location. It also informs the father and children of the migrated task of its new location.

3.3.2.3   Computing Resources

The machines are rented for utilization by tasks. If more than one application server runs on a machine several tasks may be processed simultaneously. The local broker fixes the rent. It also offers data transport and task migration services to the applications and schedules tasks to the local application servers. The local scheduling strategy may vary. In our implementation we used First-Come-First-Served scheduling. More sophisticated strategies may make use of application specific knowledge or hold local auctions.

• **Usage-based Pricing:** Similar to residential rent, which depends on the price per m^2 and space, computational rent is fixed based on two factors: price per complexity unit (PCU) and the computational effort of the task. The PCU depends on local load, demand (number of local tasks), memory capacity, and current performance as well as average global PCU and average performance. The performance of a machine is measured by the application server at runtime. This way it reflects the individual needs of the application.

• **Regulating mechanisms:** Similar to the normalized load index in conventional load distribution systems the PCU represents the machine's current state. As long as a machine is underloaded, its PCU is kept low, only depending on the global mean PCU. Low rent will attract new tasks. This, in turn, causes higher PCU. The rate of increase depends on the performance of the machine.

An overloaded workstation, on the other hand, will raise its PCU. This has different effects on local tasks. The increase in rent for tasks with moderate computational effort will be proportionally smaller than for computational intensive tasks. This results in early migrations of "large" tasks while "smaller" ones will be kept locally as long as possible.

• **Transport costs:** The costs for data transports and tasks migrations are fixed by the local brokers based on the distance between source and destination. As an important effect, tasks, exchanging data, won't move too far apart in order to keep transport costs low ("locality of communication").

Furthermore, it is possible to restrict network traffic and the degree of workstation utilization by reducing the income of the tasks. As we will see in the next sections the amount of money available to a task tree is set by the user running the application under *Dynasty*.

3.3.2.4   The Tasks

The expenditures of a task include rent for its computation and fees for services such as brokering, migrations, and data transports. In order to supply tasks with income we have adopted a mechanism similar to the sponsoring and funding procedure used in the *Spawn* project [18]. Regularly, the root task is supplied with monetary funds (income). The amount depends on the application's economical strategy and on the funds granted by the user to solve the problem. One fraction of the income is reserved to balance its future expenditures. The remaining part is distributed among the task's children. This splitting of funds is recursively repeated by the children tasks, who, in turn, supply their children with income. The funding strategy is coded into the application server.

A task may raise the income of an important subtask in order to enforce its processing. On the other hand, a task may keep more of its income to himself, e.g. for financing future brokering requests.

The root task regularly collects the current total production costs, i.e. the sum of all tasks' rent and data transport costs. This information is used to update funding of the root task, in order to ensure efficient future processing.

The money a user spends for funding the computation is subtracted from his own funds. His initial funds may be fixed according to his priority or the application's importance. As the owner of a workstation, he, in turn, earns the money that is paid by tasks utilizing his machine and using his local broker.

3.3.2.5   The Application Server

By integrating the application server as an interface between the economical system and the application code the amount of work needed for adaption is reduced and restricted to a single module. This also allows integration of applications written in different programming languages and of applications not originally designed to make use of a load distribution system. A lot of programs or modules in scientific computing are written in FORTRAN. Instead of including load distribution logic and administration code into the FORTRAN source, all extensions are put into the application server. This approach gives maximum flexibility and makes *Dynasty* a universal framework for economical based load distribution.

In order to adapt an application to run under *Dynasty* the application server has to be coded. It is provided as a C source code skeleton and has to be filled out with application specific conversions, initializations, and strategies. The adaption consists of two steps:

1. Creating the interface to *Dynasty*: Tasks are scheduled to the application server. It is therefore responsible for extracting application specific information from the task structure and for converting and transforming data. The interface also provides code to split off new children tasks.

2. Modelling the Application's economical behaviour: Strategical decisions include when and how a task is split, and when requests for brokering are issued. This way the issue of splitting and distributing the tasks is entirely hidden from the application code. Furthermore, the strategy fixes the maximum budget and the funding ratios. Economical reasoning is based on the task's funds and income, as well as local and global information. The strategy may e.g. compare the rent against the current task's income. If the difference falls below a threshold it may decide to migrate the task, spending the remaining funds on brokering.

Workstations	8	16	32	64	96
Speedup	5.13	5.59	16.29	21.50	33.82
Efficiency	0.64	0.41	0.51	0.34	0.35

Table 3.6: Speedup and efficiency for configurations with 8 up to 96 workstations

3.3.3 Performance Evaluation

In order to evaluate the performance of *Dynasty* with a practical example, we have chosen ARESO [10; 19], a parallel Finite Element solver using a divide&conquer strategy.

3.3.3.1 The ARESO application

ARESO is a Finite Element solver that uses domain decomposition and recursive sub-structuring techniques. In our case the heat distribution on a square plate is simulated. The solution on the whole domain is assembled from partial solutions of four sub-domains that can be recursively sub-structured themselves. The four subproblems can be solved in parallel. As long as the approximation after an iteration is not sufficiently precise the tree of sub-domains is refined at locations of high inaccuracy. New approximation values are calculated on the whole domain tree.

The following tests were made with up to 96 HP 9000/720 client workstations (i.e. running an application server) from two Ethernets (LANs). The first LAN consists of 8 servers with 80 MB memory each. A server is associated to 10 clients (32 MB each). The second LAN consists of 16 client workstations. A four level broker hierarchy was used with 64 and 96 client workstations, and three levels with 8, 16, and 32 client machines.

3.3.3.2 Experimental Results

The performance results are summarized in Table 3.6. The speedup increases from 5.13 on 8 workstations up to 33.82 on 96 machines. This shows that *Dynasty* is scalable with respect to the number of workstations and broker hierarchy levels. The efficiency, i.e. speedup divided by number of workstations, drops slowly from 0.64 (8 workstations) down to 0.35 (96 workstations). For this kind of divide&conquer applications these values are still satisfying and are close to the theoretically achievable efficiency, which is limited due to the huge amount of synchronization points: tasks have to wait for their parent's input data and vice versa.

Besides speedup there are several other metrics that evaluate the quality of load distribution. Table 3.7 shows the average value and standard deviation for three measures: machine utilization, PCU (rent), and the number of tasks per machine. The standard deviation represents the degree of how much the measure on one workstation deviates from the given average value of all workstations (the smaller the better).

Machine utilization is defined as the ratio of the period the machine was rent and utilized by some task (i.e. total processing time) versus total runtime. It is further presented in Fig. 3.14 for the 16 hosts configuration. There, the degree of utilization varies between 30% and 75%, with an average value of 51%.

Workstations	8	16	32	64	96
μ(Utilization)	0.72	0.51	0.55	0.34	0.32
σ(Utilization)	0.15	0.12	0.13	0.15	0.12
μ(PCU)	212	124	147	83	78
σ(PCU)	39.8	20.7	28.8	28.1	23.0
μ(#Tasks)	19.3	10.0	13.9	9.4	8.0
σ(#Task)	3.88	1.57	2.90	2.20	1.93

Table 3.7: Average value (μ) and standard deviation (σ) of machine utilization, PCU (rent) and number of independent local tasks (= load) for configurations with 8 up to 96 workstations.

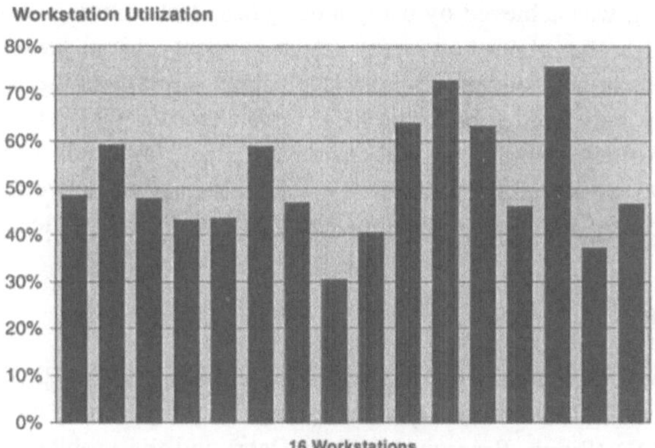

Figure 3.14: Average utilization of 16 workstations.

Fig. 3.14 indicates that, on average, all 16 machines are utilized nearly to the same degree. As is shown in Table 3.7, all standard deviations remain small with growing configuration size. This indicates that the loads on the workstations only differ slightly, even on 96 machines.

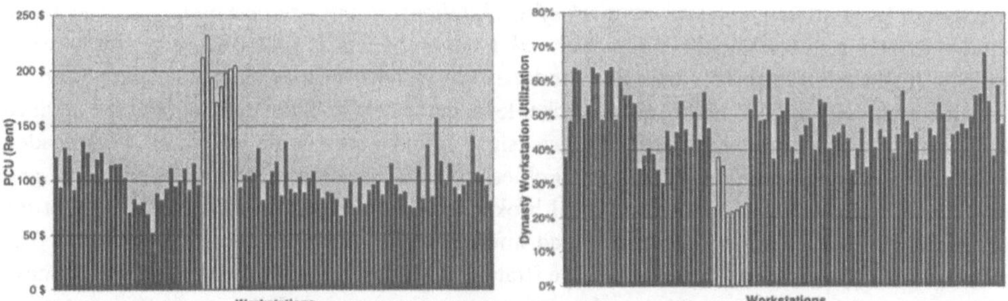

Figure 3.15: PCU (rent) and utilization of 96 workstations. Eight machines (white boxes) have additional user load.

Another key design issue under investigation is non-intrusiveness. *Dynasty* is supposed to exploit only idle workstations and, therefore, to exclude machines that are highly loaded with other jobs. We ran a dummy program on eight of 96 workstations, doing simple calculations and thus keeping the processor busy. When ARESO was started up, these eight hosts had an initial CPU load of 1.0, whereas the other 88 machines were idle. The reaction of *Dynasty* is shown in Fig. 3.15 . All of the eight machines raised their PCU (rent) above the average, thus deterring tasks with high computational effort. They only were affordable to smaller tasks. As the figure shows, their utilization by *Dynasty* dropped significantly. The rent mechanism is therefore successful in keeping a machine's utilization proportionally to its availability.

We have also investigated how much overhead *Dynasty* imposes on the application and the network. Due to the fact that the process of task splitting, i.e. generating new sub-tasks, and the process of finding appropriate target hosts overlapped, the latency of host selection was kept at a minimum. Overlapping was achieved by using a two-phase task splitting: First the application server informs the local broker about how many sub-tasks (sons) it will generate and what their budget is. While the sons are initialized the brokerage begins. Usually at the time the application server presents the fully initialized sons to the local broker their target hosts have been selected and they are moved to their destinations without delay. Brokerage took about three seconds, on average, with the 96 workstations configuration. The impact on network traffic the broker information exchange caused and the overhead of rent fixing were both neglectably small.

3.3.4 Classification

In this section *Dynasty* is classified according to the schemes of section 2. First we regard the general classification (see Table 3.8 I). *Dynasty* is *domain specific* since the computing model for applications is SPMD (Single Program Multiple Data) and the problem solving strategy is divide&conquer. This reflects also the *application oriented* intent of *Dynasty*. But since *Dynasty* is responsible for load distribution decisions and initiates the migration (motivated by raising rent) and assignment of tasks, there is also an *system oriented* intent. There is no partitioning of entities, the only load distribution function is *assignment*. Monitoring and load distribution mechansisms are integrated in the *runtime system* of *Dynasty*. The strategy, however, is implemented by the *application server*. *Dynasty* supports *location independent* problem structures.

Regarding the strategy classification (see Table 3.8 II) we adopt a *microeconomic* strategy. *Dynasty* works in an *heterogenous network of workstations of any size and topology*. The application generates a divide&conquer *tree-like task topology* which is independent of the network topology. Tasks are transfered *systemwide* and they are assigned *non-preemptive*. Load information is hierarchically aggregated from one broker level to the next and thus only obtainable within a *short range*. The information scope is *complete*, since global system wide information is spreaded from top of the broker hierarchy down to all other brokers. Decision making is *hierarchical* and involved brokers *cooperate*. Only two local brokers participate load distribution activity at the same time, i.e. participation is *partial*. Load information is spreaded among the brokers. The sender initiates load distribution activity. The strategy is adjustable since application servers have access to rent and price information and can adopt to an appropriate behaviour. The strategy *is concerned about costs and stability* since tasks have only limited funds to cover data transfer and migration costs.

I. General

Objectives	
domain specific	D&C, SPMD
intent	both
function	assignment

Integration Level	
monitoring	runtime system
strategy	application
mechanisms	runtime system
compiler support	none

Structure	
problem structure	location independent
subproblems	computations
workspaces	data (tasks)
targets	appl. server (processes)

II. Strategy

System Model	
Model Flavour	microeconomic
Target Topology	het., NOWs
Entity Topology	tree-like

Transfer Model	
Transfer Space	systemwide
Transfer Policy	non-preemptive

Information Exchange	
Information Space	short range
Information Scope	complete

Coordination	
Decision Structure	hierarchical
Decision Mode	cooperative
Participation	partial

Algorithm	
Decision process	dynamic
Initiation	sender
Adaptivity	adjustable
Cost Sensitivity	high
Stability control	garantueed

Table 3.8: General and Strategy Classification

III. Load Model

Load Index Properties	
index dimension	multi dim.
index type	appl./CPU,MEM
aggregation time	inst./compr.
aggregation space	hybrid

Load Value Composition	
index combination	weighted linear
weight selection	pre-runtime

Model Usage Policies	
index measurement	periodically
index propagation	adaptive
model adaptivity	fixed

IV. Migration Mechanism

Migrant	tasks
Set Size	limited
Heterogeneity	yes
Initiation	immediate
Pre Transfer Media	—
Compression	no
Transfer Policy	complete
Residual Dependency	none
Mig. Transparency	yes
Code Transparency	yes

Table 3.9: Load Model and Migration Mechanism Classification

The load model (see Table 3.9 III) uses a *multi dimensional* load index, which is *application and system defined*, since the application measures its subjective processing speed while the system provides information about available memory, CPU load, number of interactive users etc. From this index the single dimensional rent value is calculated. There is either *instantenous* or *comprehensive* aggregation time for the load index. Measurements of the load index cover *application and system* defined objects. The load index and rent is measured and propagated *periodically* by the broker system. The load index *does not change* during runtime.

The last classification scheme (see Table 3.9 IV) concerns the used migration mechanisms. *Dynasty* migrates *tasks*, i.e problem specifications. The *migration* is done by *Dynasty* and is *not obvious* to the application. Code transparency is given, since the application uses an application server as its transparent interface to the system. In a single migration only one task can be migrated, i.e. migration set size is *limited*. Migration requests are handled *immediately*. Tasks are *residual independent*. It is possible to migrate tasks between *heterogenous* nodes. For migration tasks are *not compressed* and they are *completely* transferred.

3.3.5 Conclusions

In this chapter we introduced the concepts, implementation, and evaluation of *Dynasty*, an economic-based dynamic load distribution system for SPMD parallel applications in large heterogeneous computer networks. Its goal is to distribute tasks to idle workstations with migration on demand. This is accomplished by implementing a computational economy based on a price mechanism for regulating supply and demand. Workstations fix local execution costs (= rent) depending on current performance, demand, global mean rent, and a task's computational effort. Rent, as an abstraction for conventional load indices, proves to be an elegant mechanism for dealing with machines that differ in performance and capability. The hierarchical organized broker system assigns tasks to machines. A task, demanding migration, specifies its target host requirements and selection preferences. The selection range depends on the task's budget and current brokerage. Thus, by allocating funds to tasks adequately, selection range and migration distance is proportionally to a task's importance.

We evaluated *Dynasty* by adapting the Finite Element solver ARESO. The adaption required only minor changes in the original application code. All ARESO dependent code concerning economic reasoning and data conversion was placed in the application server, the interface between the application and *Dynasty*. The measurements using up to 96 workstations produced satisfying speedups, proving *Dynasty*'s scalability with respect to network size. The tests showed only slight deviations in the workstations' average load (number of independent local tasks). Furthermore, due to our rent mechanism, the utilization of the workstations was inversely proportional to the load caused by other programs and users, yielding fair load distribution. The experiments also proved that the chosen economical reasoning, coded into the ARESO application server, could be re-used for different problems and network configurations without any modifications.

3.3.6 Future Work

The evaluations showed that *Dynasty* is a valuable platform for load distributing. Besides using larger networks and other divide&conquer applications, we will consider more general forms of task dependencies. Currently, we are working on the adaption of *Dynasty* to a parallel functional language [4] that makes use of runtime-fixed task dependencies.

Other issues of future research are robustness, security, and the development of more sophisticated broker and application server strategies. Smart brokers may accumulate knowledge about application behaviour. They could learn about periods in which workstations are idle or are used interactively, resulting in more precise and reliable rent forecasting.

References

[1] M. BACKSCHAT. Entwicklung eines hierarchisch organisierten Maklersystems zur dynamischen Lastverteilung von Divide&Conquer-artigen Problemen. Master's thesis, Technische Universität München, Institut für Informatik V, http://www5.informatik.tu-muenchen.de/publikat/diplarb.html, 1995.

[2] T. CASAVANT. A Taxonomy of Scheduling in General-Purpose Distributed Computing Systems. *IEEE Trans. Softw. Eng.*, 14(2):141–154, February 1988.

[3] D. CLARKE and B. TANGNEY. Microeconomic Theory Applied To Distributed Systems. Technical Report TCD–CS–93–30, Distributed Systems Group, Department of Computer Science, University of Dublin, 1993.

[4] R. EBNER and A. PFAFFINGER. Transformation of Functional Programs into Data Flow Graphs Implemented with PVM. In *3rd Euro PVM Users' Group Meeting*, Munich, October 1996. Springer.

[5] D. FERGUSON. Microeconomic Algorithms for Load Balancing in Distributed Computer Systems. In *IEEE Int. Conf. on Distributed Computer Systems*, pages 491–499, 1988.

[6] A. GEIST, A. BEGUELIN, J. DONGARRA, W. JIANG, R. MANCHEK and V. SUNDERAM. *PVM: Parallel Virtual Machine — A Users' Guide and Tutorial for Networked Parallel Computing*. Scientific and Engineering Computation. The MIT Press, Cambridge, MA, 1994.

[7] M. HAILPERIN. *Load Balancing using Time Series Analysis for Soft Real Time Systems with Statistically Periodic Loads*. Ph.d. thesis, Stanford University, December 1993.

[8] H. U. HEISS. Dynamic Decentralized Load Balancing: The Particles Approach. In L. GÜN, L. ONVURAL and E. GELENBE, editors, *Int. Symp. on Computer and Information Sciences VIII*, Istanbul, November 1993.

[9] B. HUBERMAN and T. HOGG. The Behaviour Of Computational Ecologies. In B. HUBERMAN, editor, *The Ecology of Computation*, pages 77–115. Elsevier Science Publishers (North Holland), 1988.

[10] R. HÜTTL and M. SCHNEIDER. Parallel Adaptive Numerical Simulation. SFB report 342/01/94 A, Institut für Informatik, Technische Universität München, 1994.

[11] H. KUCHEN and A. WAGENER. Comparison of Dynamic Load Balancing Strategies. Aachener Informations-Bericht 90–5, RWTH Aachen, Lehrstuhl für Informatik II, 1990.

[12] M. LITZKOW. Condor – A Hunter of Idle Workstations. In *Proceedings of the 8th International Conference on Distributed Computing Systems*, pages 104–111, San Jose, California, 1988.

[13] T. W. MALONE. Enterprise: A market-like Task Scheduler for Distributed Computing Environments. In B. HUBERMAN, editor, *The Ecology of Computation*, pages 177–205. Elsevier Science Publishers (North Holland), 1988.

[14] B. MCMILLIN and H. CLARK. DAWGS — A Distributed Compute Server Utilizing Idle Workstations. *Journal of PDC*, 14:175–186, 1992.

[15] J. C. PASQUALE. *Intelligent Decentralized Control in Large Distributed Computer Systems*. Ph.d. thesis, Computer Science Division, University of California, Berkeley, 1988. Report No. UCB/CSD 88/422.

[16] R. POLLAK. A Hierarchical Load Balancing Environment for Parallel and Distributed Supercomputer. In *PDSC'95, Int. Symp. on Parallel and Distributed Supercomputing*, Fukuoka, Japan, 1995.

[17] J. ROSENSCHEIN and M. GENESERETH. Deals Among Rational Agents. In B. HUBERMAN, editor, *The Ecology of Computation*, pages 117–132. Elsevier Science Publishers (North Holland), 1988.

[18] C. A. WALDSPURGER. Spawn: A Distributed Computational Economy. *IEEE Trans. Softw. Eng.*, 18(2):103–116, 1992.

[19] C. ZENGER, W. HUBER, R. HÜTTL and M. SCHNEIDER. Distributed numerical Simulation on Workstation Networks. In *Symp. on HPSC High Performance Scientific Computing*, Munich, June 1994.

[20] S. ZHOU. UTOPIA: Load Sharing in Large-Scale Heterogeneous Distributed Systems. Technical Report CSRI–257, Computer Systems Research Institute, University of Toronto, April 1992.

3.4 ALDY - an Adaptive Load Distribution System

by Claudia Gold and Thomas Schnekenburger

3.4.1 Introduction

In distributed systems where several users can simultaneously access the same resources of the system additional load on the individual resources caused by other applications cannot be predicted. Therefore a load distribution system is needed for the execution of parallel programs which allows to *adapt* the workload distribution to the actual resource usage of other applications during run-time.

Two different methods of load distribution are distinguished: System based load distribution defines distribution objects (entities) implicitely and assigns them automatically to distribution units (targets). Moreover in system based load distribution systems the distribution mechanisms are usually transparent for the application. Using application based load distribution, the application programmer defines entities on his own and is also responsible for the assignment of entities to targets. This method is more flexible than the system based method, because real application objects are taken into account for distribution and the granularity of the distribution objects can be determined by the application. The disadvantage, however, is the fact, that the application has to be changed for the insertion of load distribution mechanisms.

This section introduces the Adaptive Load Distribution SYstem ALDY. It is a hybrid between system based and application based load distribution and combines the advantages of both. The programmer defines his own distribution objects but the assignment and migration of these distribution objects to targets during run-time are controlled by ALDY.

In Sect. 3.4.2 the concepts and design of ALDY are surveyed. Section 3.4.3 shows a PVM application as an example to demonstrate the integration of ALDY. Section 3.4.4 refers to related work. In Sect. 3.4.5 the classification schemes of Sect. 2 are applied for ALDY. Section 3.4.6 contains a conclusion and gives an overview over future work concerning ALDY.

3.4.2 Concepts and Design

ALDY is not a complete environment for developing parallel programs. It is a library providing specific functions supporting the implementation of load distribution. In this section the concepts and the design of ALDY are discussed.

3.4.2.1 Design Objectives

Portability is a very important factor for the design of ALDY. There are four concepts followed to reach portability:

The first concept is to be independent from a specific implementation of the distribution objects. Therefore the application programmer maps his own objects to virtual objects of the ALDY load distribution model. The different kinds of virtual objects are listed in Sect. 3.4.2.2.

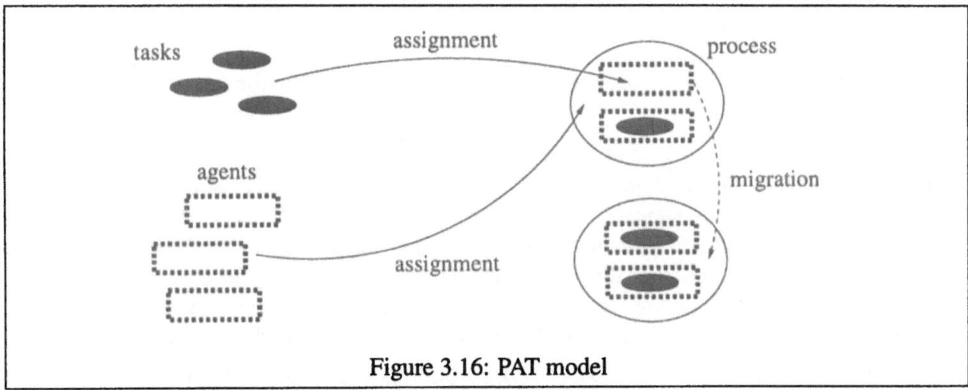

Figure 3.16: PAT model

Secondly, ALDY provides a collection of different strategies, from which for individual applications the most suitable one can be chosen. It is not necessary to change the application program for testing another strategy. The new strategy can be chosen at run-time without recompilation.

To be independent from system parameters and consequently from individual resources, the ALDY strategies use only information about the states of virtual objects (see Sect. 3.4.2.3). System states like the load of hosts are not regarded.

At last ALDY is not fixed to a specific communication and synchronization platform. It only provides a kernel system for communication and synchronization. ALDY realizes communication and synchronization with call back functions which have to be implemented by the application programmer. By implementing these routines the programmer decides on which system the communication should be realized. ALDY uses the call back functions also for its internal communication.

3.4.2.2 Load Distribution Model

A load distribution model defines distribution objects and distribution units. The load distribution model of ALDY is called the PAT (**Process, Agent, Task**) model (cf. Figure 3.16). It consists of three classes of virtual objects:

- **Virtual processes** correspond to the distribution units. The goal of ALDY is to assign workload to virtual processes. Therefore, they represent distributed resources of the system. In most cases virtual processes are real application processes.
- **Virtual agents** are objects that are assigned preemptively to virtual processes. Therefore, they can be migrated dynamically to another process. Several agents may be assigned simultaneously to the same process.
- **Virtual tasks** are objects that are assigned non-preemptively to agents. This saves costs for migrating agents, since agents can only be migrated if they are not processing a task. Tasks either represent messages which are assigned to agents or operations which are processed by agents. Each agent processes one task after another. There is no parallelism within agents.

All the above described virtual objects are represented by integer identifiers.

3.4.2.3 Library Interface

The interface between the application and ALDY consists of functions for describing virtual objects and their attributes. Some of the functions of the C-library interface are described below.

To realize efficient protocols for initialization, termination and other distributed events, ALDY requires all virtual processes to be arranged in a ring. This ring is used for the internal communication of ALDY. It is not automatically built by ALDY. Instead each virtual process has to call the library function `ALDY_Init` to form and initialize the ring. That means the application programmer is responsible for constructing the ring. Thus he has the possibility to organize the virtual processes in the ring corresponding to the demands of the application. For example, two virtual processes locating agents which communicate intensively should be neighbours in the ring.

A call to the function `ALDY_DefineObject` informs ALDY about the identifiers for the used tasks and agents. Since ALDY allows to specify directives for virtual objects which determine where an object shall be assigned (see Sect. 3.4.2.4), the application programmer must inform ALDY when a virtual objects is completely specified. This is done with a call to the function `ALDY_ReadyObject`. ALDY can then assign the object.

ALDY does not use system information for its strategies. The number of active agents within a process serves as load index and an agent is active, when it executes a time consuming part of the application. Since ALDY does not know the implementation of the object, which is represented by an agent, the application programmer has to inform ALDY when an agent is active. This is done with the functions `ALDY_StartAction` respectively `ALDY_EndAction`. Based on the definition of active agents a process is defined as active, if at least one agent is active. If now e.g. the load on a processor increases because another application is started, the time a process is active for processing a task on this processor also increases. This will cause the migration of agents from that processor to less loaded processors.

As long as a task is assigned to an agent, the agent cannot be migrated (see Sect. 3.4.2.2). The call of `ALDY_NextTask` terminates the current task and requests a new task from ALDY. That means the agent to which the terminated task was assigned can now be migrated until a new task is assigned.

Other functions of the library interface are used by ALDY to inform the application about load distribution mechanisms (see Sect. 3.4.2.5).

3.4.2.4 Directives

An important facility of ALDY is the possibility to specify directives for the assignment of virtual objects (library function `ALDY_AddDirective`). The task-agent-assignment directive e.g. specifies a set of agents to which a task may be assigned. This directive can be used by applications to define the responsibility of agents for processing a specific task.

Another directive that should be mentioned allows the definition of neighbourship among agents. In most cases an agent only communicates with a few other agents. With the agent-neighbourship directive tightly coupled communicating agents can be defined as neighbours. This reduces communication amount, since ALDY tries to assign neighbouring agents to the same process.

3.4.2.5 Mechanisms for Load Distribution in ALDY

ALDY works with virtual objects and does not know the implementation of the real application objects. Therefore ALDY cannot migrate the real application objects on its own. It only determines when and where an agent should be assigned or migrated and calls then the corresponding call back functions `ALDY_CB_AssignAgent` respectively `ALDY_CB_MigrateAgent`. That means the mechanisms for assigning and migrating application objects are implemented by the application programmer.

Further communication between the processes is needed for load distribution. On one hand ALDY has to get information about the load of a process to decide whether an agent should be migrated or not. On the other hand tasks have to be sent to agents, which are migratable. Usually message-passing between migratable objects is very complex and often causes errors in parallel programs. Therefore ALDY takes the responsibility for delivering messages (in form of tasks) correctly to agents. But since the implementation of tasks is not known to ALDY, the passing of tasks is done via call back functions. These call back functions are also used to send internal ALDY messages. The relevant call back functions are listed below.

`ALDY_CB_Receive`	–	receive message for ALDY
`ALDY_CB_ObjectInfo`	–	pass received object to application
`ALDY_CB_SendInfo`	–	send ALDY information to another process
`ALDY_CB_SendObjectInfo`	–	send object and ALDY information to another process

For the implementation of these functions the same message-passing platform can be used as the one on which the application is based. The following example (taken from [8]) illustrates an implementation of a call back function with PVM.

Example: (Implementation of the call back function `ALDY_CB_SendObjectInfo` with PVM3)

`ALDY_CB_SendObjectInfo` has to send a message to process `pid`. The message contains object `ob` and internal ALDY information. Generally, internal ALDY messages are coded as integer arrays, because this alleviates marshalling. Here the array starts at address `addr` and has the length `n`. Since ALDY does not know the real object data, these have to be defined by the application. The return value of `ALDY_CB_SendObjectInfo` is the number of additional bytes needed for sending object `ob`.

```
int ALDY_CB_SendObjectInfo(int ob, int pid, int * addr, int n)
{
  pvm_initsend(PvmDataDefault);            /* initialize send-buffer */
  pvm_pkint(&n,1,1);                       /* put length of ALDY-message */
  pvm_pkint(addr,n,1);                     /* put ALDY-internals */

  /* assuming object has 1000 bytes starting at address obAddr */
  pvm_pkbyte(obAddr,1000,1);              /* put object into the buffer */

  /* assuming that the array tids contains pvm-identifiers */
  /* send ALDY message, tag 999 is used to mark ALDY-messages */
  pvm_send(tids[pid-1], 999);

  return 1000;                            /* return number of bytes of object */
}
```

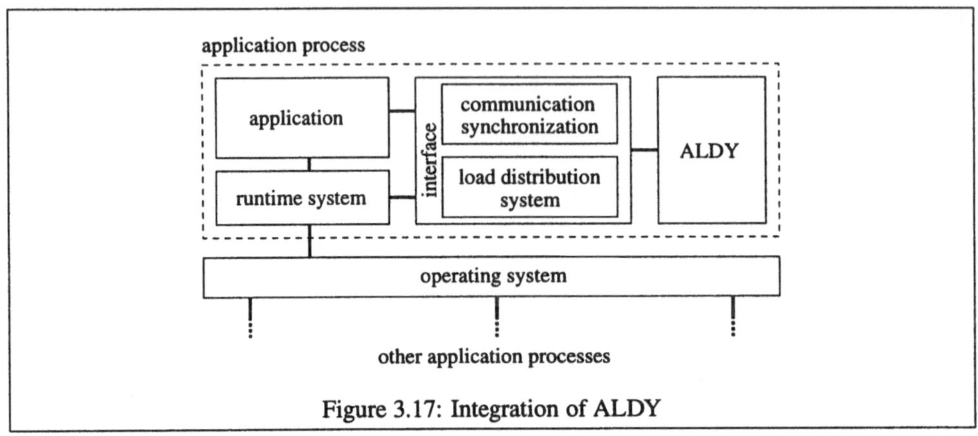

Figure 3.17: Integration of ALDY

3.4.2.6   Load Distribution Strategies

As mentioned in Sect. 3.4.2.1, ALDY separates the application program from the load distribution strategy. ALDY uses a generic load distribution concept that allows to integrate a collection of several parameterized specific load distribution strategies.

To realize this concept, ALDY uses an internal and an external strategy. The internal strategy is responsible for efficient global information management. It provides information about states, directives, and actual locations of virtual objects. This information is used by the external strategy, which has to decide whether and where an agent should be assigned or migrated. For this decision a receiver-initiated strategy (see [3]) is currently used.

The criteria on which the external strategy makes its decisions are defined in a special parameter file that is read at the beginning of the initialization of ALDY. The individual strategy is therefore not part of the library interface.

3.4.2.7   Programming with ALDY

Figure 3.17 illustrates how ALDY is integrated into parallel programs. Load distribution with ALDY is performed on application level. That means, from the operating system's point of view, ALDY and the application form together a single parallel program. The ALDY system is linked to each process of the parallel program. The application interacts only locally with the corresponding ALDY instance. The library interface (see Sect. 3.4.2.3) connects the application and ALDY.

Steps for programming a parallel algorithm with ALDY

The first step is the mapping of application objects to virtual ALDY objects (see Sect. 3.4.2.2). This can be a very difficult job. For example an attempt to integrate ALDY into a numerical application [6], called NSFLEX, failed unfortunately. It failed, because the FORTRAN 90 code of NSFLEX, which was frequently changed and enlarged in the past, is very complex. Therefore the code was extremely hard to understand and to modify for the integration of ALDY. The problem of mapping application objects to virtual objects is to find suitable dynamic data structures for the original static data structures. Migratable parts must be separated from the original static data

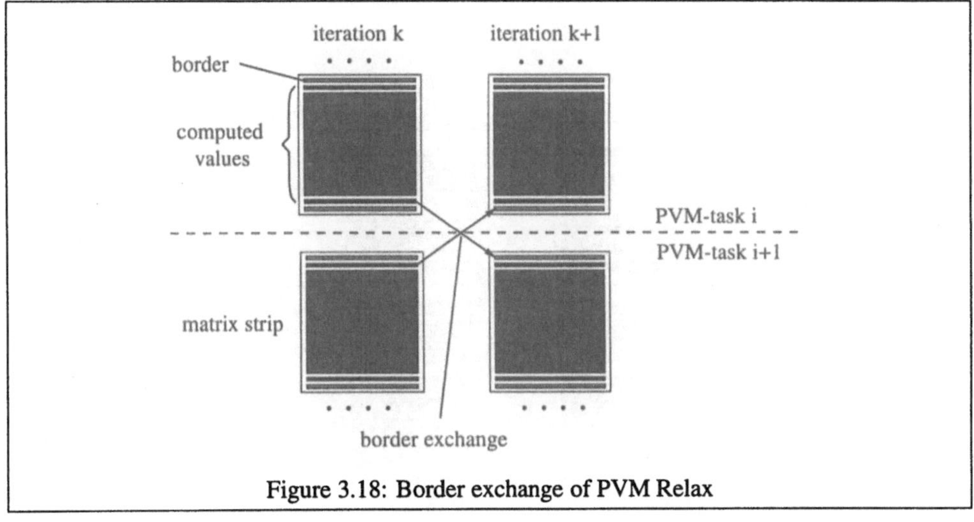

Figure 3.18: Border exchange of PVM Relax

structures. Also the granularity of the migratable parts (agents) must be considered because the granularity determines the quality of load balance that can be reached. If the number of virtual processes is identical with the original number of processes, finer granularity does not increase communication amount, since communication of agents within a process is local (cf. dashed lines in Figure 3.19). But if the size of agents is chosen very small, the number of migrations will grow, which inreases the overhead caused by migrations.

In the next step the virtual objects must be defined and supplied with attributes. The task of the application programmer is to find unique integer identifiers for all virtual objects. Also the corresponding calls to the ALDY library functions must be inserted into the application program.

At last the call back functions are implemented using the communication system which is available to the application programmer.

3.4.3 Example

The example PVM Relax [7] demonstrates how ALDY can be integrated in existing applications. PVM Relax is a parallel implementation of a Relaxation algorithm using PVM. Given is a grid of metal with a constant temperature distribution around the grid. The problem is to determine the temperature distribution that will be reached inside the grid after a while.

To solve this problem for each point inside the grid the average temperature value of all surrounding points is computed for a number of iterations. The number of iterations can be reduced by using a numerical method called Relaxation, that also led to the name PVM Relax.

For the numerical solution of this problem a matrix is used as basic data structure, where each component of the matrix contains the temperature of a point of the metal grid.

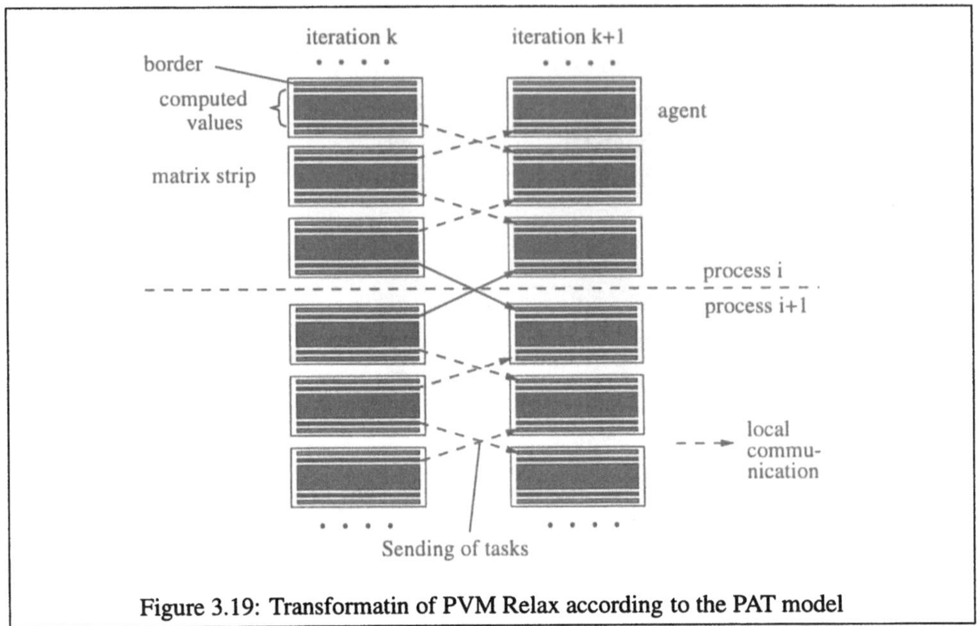

Figure 3.19: Transformatin of PVM Relax according to the PAT model

3.4.3.1 Parallelizing with PVM

The Relaxation algorithm can be parallelized by partitioning the matrix into several strips (for stripwise decomposition see [5]), where each strip keeps some adjacent lines of the original matrix. For the computation of each strip one process is responsible.

Before a new iteration can be computed, each strip needs the data of the previous iteration from the neighbouring strips for its lower and upper border line (see Figure 3.18). That means, before each iteration a PVM-task has to wait for the border temperature values of its strip with pvm_recv. After each iteration the new computed values of the border lines are sent to the neighbouring PVM-tasks with pvm_send. The border values of each strip are only read for the computation of the inner values. They are not changed while computing the iteration.

3.4.3.2 The Integration of ALDY

The pure PVM solution does not offer the possibility to migrate workload from one PVM-task to another. The execution of PVM Relax can be slowed down, if only one PVM-task is located on a slow host. This can be avoided by using ALDY.

For the integration of ALDY, PVM Relax has to be transformed according to the PAT model (see Figure 3.19). Since workload cannont be migrated among PVM-tasks virtual processes are introduced which represent PVM-tasks. The idea for load distribution is now to divide the original matrix strips again into smaller ones and to assign several strips to one process. Load balance will then be achieved by migrating strips from overloaded to underloaded processes. Therefore the strips are represented by agents and several agents are located on a virtual process. In the pure PVM solution the border exchange of each layer is realized with PVM-messages. Since now these data have to be sent between migratable agents, virtual tasks are used for sending borders. An additional advantage for using tasks is the fact, that during processing of tasks no agent migration is possible. This reduces overhead caused by stopping an agent anywhere during the computation of an iteration. Furthermore, the implementation of an agent is made easier because less information is needed if an agent can only be transferred at a few exact defined execution points.

After the mapping of real application objects to virtual objects has been performed, the calls to the ALDY library functions are inserted and the call back functions have to be implemented (see Sect. 3.4.2.5). The communication is again done by PVM.

Since agents which are processing neighbouring strips communicate intensively, these are defined as neighbours (see Sect. 3.4.2.4).

3.4.3.3   Results

With the integration of ALDY, the workload of the different PVM-tasks can dynamically be adapted to the varying workload of the application and to the actual speed of individual nodes. All difficult parts concerning the migration of objects are automatically provided by the ALDY system. Only the call back routines for sending and receiving objects have to be implemented by the application programmer.

Another advantage of ALDY is the fact, that different load distribution strategies for PVM Relax can easily be exchanged and tested without any modification of the application code for finding e.g. the most appropriate strategy.

Experiments [6] concerning the efficiency of the integration of ALDY into PVM Relax were conducted on an Ethernet network of nine SUN SPARCstations ELC. Each workstation hold one process to which eight agents were assigned at the beginning of program execution. The workload of PVM Relax is equally distributed among the processes, because each strip is representing a partial matrix of the same size and at program start the same number of agents is assigned to each process. To simulate unbalanced workload and to force agent migration for the experiments, the values of the different lines of the matrix were computed a different number of times for each iteration according to the sinus function of Figure 3.20 . For example the lines 100 and 300 were computed five times and the line 200 was computed ten times. The used matrix consisted of 400 lines and rows and 300 iterations were made. The run-time of the pure PVM solution was between six and nine minutes. Figure 3.21 illustrates the results of some of the experiments, which were made with different parameter settings.

- The first two bars show the overhead caused by ALDY which amounts to about 5%. This means that the pure PVM solution is about 5% faster than the solution with ALDY, if no distribution of agents is allowed.
- The remaining bars represent the run-time values which were obtained using a receiver-initiated strategy with different parameter settings. The parameter **R** determines the value of the load index below which a process requests a new agent. **S** defines the threshold which must be exceeded for delivering an agent. The best improvement (last bar of Figure 3.21) that was reached in all experiments with the above described experiment structure resulted in about 30%. This means that the solution using the load distribution of ALDY

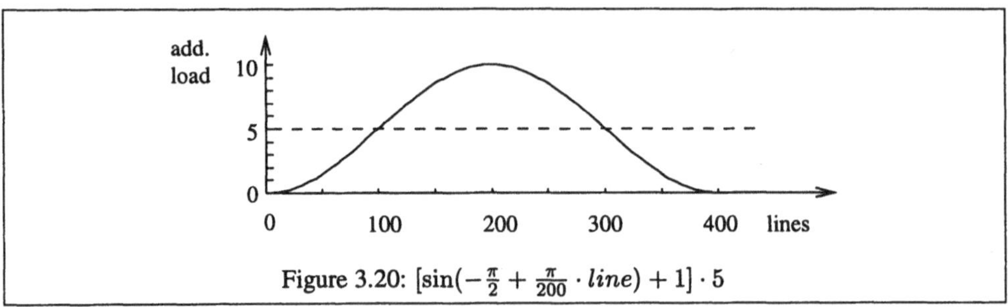

Figure 3.20: $[\sin(-\frac{\pi}{2} + \frac{\pi}{200} \cdot line) + 1] \cdot 5$

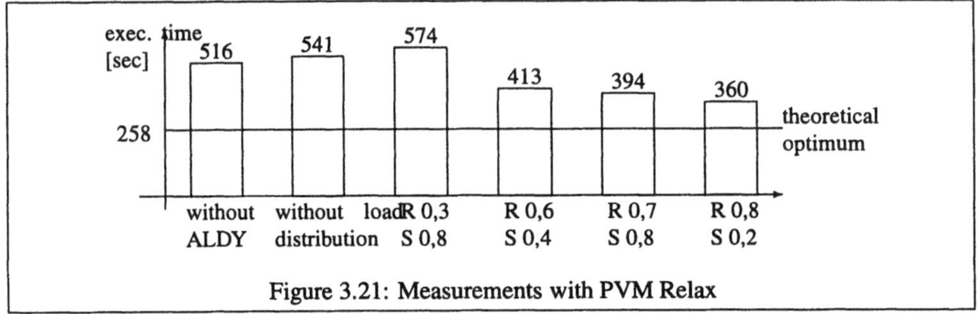

Figure 3.21: Measurements with PVM Relax

was about 30% faster than the solution without ALDY. Depending on the sinus function (see Figure 3.20), which was chosen to unbalance the workload, the maximal improvement that could be reached theoretically is 50%. This value can be explained by the fact, that the average number of times each line is computed is five (dashed line in Figure 3.20) which is exactly half of the maximal number of times. Consequently we obtained 60% of the maximal improvement by integrating ALDY.

3.4.4 Related Work

There are several projects that also treat adaptive load distribution using real application objects as distribution objects. Like ALDY all these systems are realized by libraries which are linked to the application program.

CASAS *et al.* [2] describe three load migration systems for PVM. The systems MPVM (Migratable PVM) and UPVM follow pure system based load distribution concepts. However, ADM (Adaptive Data Movement) is a pure application based load distribution system. CASAS *et al.* report an overhead of about 23% that is caused by the integration of ADM into an application while the integration of ALDY only causes an overhead of about 5% (see Sect. 3.4.3.3, page 111).

THE PARFORM [1] is a platform for parallel programming in distributed systems. It allows in addition to adaptive load balancing heterogenous partitioning with respect to the load situation on the different hosts at startup. Heterogenous partitioning is also supported by ALDY, since the appropriate virtual process for the assigment of an agent is chosen according to the load distribution strategy. That means the number of agents on the different processes can vary at startup.

In THE PARFORM communication is fixed to sockets using TCP/UDP protocols and there is no possibility for the user to influence the load distribution strategy.

Dynamo [12] is a tool that supports the development of parallel programs which perform adaptive load distribution. However Dynamo assumes independent work units (entities) which are assigned non-preemptive to targets. That means they cannot be migrated after their processing had started. Dynamo offers the application programmer an interface for developing his own distribution strategies. ALDY does not provide such an interface. For integrating a new strategy the application programmer has to change the source code of ALDY. Communication in Dynamo is done with the PICL communication library [4].

3.4.5 Classification

This section classifies ALDY according to the schemes of Sect. 2.

First we regard the general classification (see Table 3.10, page 114). ALDY is *not domain specific* since agents are application defined. This reflects also the *application oriented* intent of ALDY. There is no partitioning of entities, the only load distribution function is *assignment*. Monitoring and strategies are integrated into the *runtime system* ALDY. The load distribution mechansisms, however, are implemented via call back functions by the *application*. For load distribution ALDY does *not* offer specific language constructs which are translated by a compiler. But SCHNEKENBURGER [10] has already given some thought to compiler support. The runtime system ALDY is linked to each application process (see Sect. 3.4.2.7). ALDY supports *location dependent* problem structures with *object accesses*, i.e. accesses to agents, as subproblems. Workspaces are represented by *objects* (agents) and targets are virtual *processes*.

Regarding the strategy classification (see Table 3.10, page 114) we restrict us to the *receiver initiated* strategy, which is the only currently implemented strategy. ALDY assumes a *fully* connected target topoloy, which may be *heterogenous*. The application can specify neighbourships among agents and therefore the strategy can - *dependent* on the *application* - take different kinds of entity topologies into account. Agents are only transferred to processes, on which *neighbouring* agents are located, and they are assigned *preemptively*. For migrating an agent load state information is asked from processes which keep *neighbouring* agents. Therefore the information scope is *partial*. Decision making is *distributed* (each process can decide) and the involved processes *cooperate*. Only two processes participate in load distribution activity at the same time, i.e. participation is *partial*. Load information is requested *dynamically* during run-time. The model flavour already says that the *receiver* initiates load distribution activity, the receiving process asks for agents if the load value is under a specific threshold. The strategy is *fixed* and cost sensitivity is *partial*, since the entitiy topology is always taken into account. Also stability control is *partial*, because the thresholds specified in the parameter file influence the migration decisions.

The load model (see Table 3.4.5, page 115) uses a *single dimensional* load index, which is *application* defined, since the application decides when an agent is active or not (cf. Sect. 3.4.2.3, page 106). There is either *instantenous* or *comprehensive* aggregation time for the load index possible. This can be chosen in a parameter file, which is read at program start. Measurements of the load index only cover states of agents, *application* defined objects. The load index is measured whenever an ALDY library function is called, which *depends* among other things on the time agents are active. That means the initiation of index measurement is influenced by the *application*. Load index propagation is performed *on demand*.

<table>
<tr><td colspan="2">I. General</td><td colspan="2">II. Strategy</td></tr>
</table>

I. General			II. Strategy	

I. General

Objectives	
domain specific	no
intent	application adaptive
function	assignment
Integration Level	
monitoring	runtime system
strategy	runtime system
mechanisms	application
compiler support	none
Structure	
problem structure	location dependent
subproblems	object access
workspaces	objects
targets	processes

II. Strategy

System Model	
Model Flavour	receiver initiated
Target Topology	het. fully connected
Entity Topology	application dependent
Transfer Model	
Transfer Space	neighbourhood
Transfer Policy	preemptive
Information Exchange	
Information Space	neighbourhood
Information Scope	partial
Coordination	
Decision Structure	distributed
Decision Mode	cooperative
Participation	partial
Algorithm	
Decision Process	dynamic
Initiation	receiver
Adaptivity	fixed
Cost Sensitivity	partial
Stability Control	partial

Table 3.10: General and Strategy Classification

The last classification scheme (see Table 3.4.5, page 115) considers the used migration mechanisms. ALDY migrates *objects* – exactly agents. In a single migration only one agent can be migrated, i.e. migration set size is *limited*. Agents can be migrated between *heterogenous* nodes. Migration requests are handled *delayed*, after the current task was processed by the agent. Whether agents are compressed for migration or not, is *application dependent*, since the application programmer implements the corresponding call back functions. Agents are *completely* transferred and they are *residual independent*. The migration is done by ALDY and therefore *transparent* for the application. Code transparency is *not* given, since the call back functions have to be implemented for each application.

3.4.6   Conclusion and Future Work

One basic and central concept, in which ALDY differs from earlier load distribution systems, is the use of virtual objects according to the PAT model. This allows the use of many different kinds of distribution objects. If later versions of PVM would e.g. provide threads, these could also be mapped to agents. Other new concepts of ALDY are the implementation of data exchange on top of the application's communication mechanism via call back functions and the collection of load distribution strategies.

III. Load Model

Load Index Properties	
index dimension	single dim.
index type	application
aggregation time	inst./compr.
aggregation space	application
Load Value Composition	
index combination	—
weight selection	—
Model Usage Policies	
index measurement	application dependent
index propagation	on demand
model adaptivity	fixed

IV. Migration Mechanism

Migrant	object
Set Size	limited
Heterogeneity	yes
Initiation	delayed
Pre Transfer Media	—
Compression	application dependent
Transfer Pol.	complete
Res. Dep.	none
Mig. Transp.	yes
Code Transp.	no

Table 3.11: Load Model and Migration Mechanism Classification

Currently ALDY is implemented as a prototype. There are still some functions missing. Also this prototype includes only one strategy out of a variety of strategies. The current strategy is a receiver-initiated one. This strategy is very efficient, but it would be interesting to test a corresponding sender-initiated strategy. Strategies with e.g. microeconomic or physical analogies should also be considered.

Moreover we are looking for further applications to test ALDY. We need more information about when and where it is useful to apply ALDY. Also some more measurements about the efficiency of ALDY are interesting.

References

[1] C. H. CAP and V. STRUMPEN. Efficient parallel computing in distributed workstation environments. *Parallel Computing*, 19:1221–1234, 1993.

[2] J. CASAS, R. KONURU, S. W. OTTO, R. PROUTY and J. WALPOLE. Adaptive Load Migration systems for PVM. In *Supercomputing'94*, pages 390–399, 1994.

[3] D. L. EAGER, E. D. LAZOWSKA and J. ZAHORJAN. A Comparison of Receiver-Initiated and Sender-Initiated Adaptive Load Sharing. *Performance Evaluation*, 6:53–68, 1986.

[4] G. A. GEIST, M. T. HEATH, B. W. PEYTON and P. H. WORLEY. PICL – A Portable Instrumented Communication Library. Technical Report TM-11130, Oak Ridge National Laboratory, 1990.

[5] R. V. HANXLEDEN and L. R. SCOTT. Load Balancing on Message Passing Architectures. *Journal of Parallel and Distributed Computing*, 13:312–324, 1991.

[6] M. KORNDÖRFER. Anwendung einer Lastverteilungsbibliothek für ein paralleles numerisches Programm. Master's thesis, Technische Universität München, Institut für Informatik, November 1995.

[7] C. PLEIER and G. STELLNER. PVM in der Praxis: Temperaturverteilung parallel berechnet. *iX – Multiuser Multitasking Magazin*, pages 186–191, December 1994.

[8] T. SCHNEKENBURGER. The ALDY Load Distribution System. Technical Report SFB 342/11/95 A, Technische Universität München, May 1995.

[9] T. SCHNEKENBURGER. Supporting Application Integrated Load Distribution for Message Passing Programs. In *ParCo (Parallel Computing)*, pages 439–446, Gent, Belgium, September 1995. Elsevier.

[10] T. SCHNEKENBURGER. Cooperating Agents: Language Support and Load Distribution. In *11 th Int. Parallel Processing Symposium, Workshop on High-Level Parallel Programming Models and Supportive Environments, Geneva, Switzerland*. IEEE, 1997.

[11] V. S. SUNDERAM, G. A. GEIST, J. DONGARRA and R. MANCHEK. The PVM concurrent computing system: Evolution, experiences, and trends. *Parallel Computing*, 20(4):531–546, 1994.

[12] E. TÄRNVIK. Dynamo - a portable tool for dynamic load balancing on distributed memory multicomputers. *Concurrency: Practice and Experience*, 6(8):613–639, December 1994.

4 Applications

In the previous chapter several systems that offer a different approach to load distribution have been described. This chapter now deals with application based solutions to distribute the workload. Again the presentation usees the terminology defined in chapter 2.

In the first section the use of an Interface Definition Language to address the problem of load distribution is explained. Section 4.2 shows how load distribution techniques can be applied in a parallel theorem prover. In section 4.3 it is described how load distribution methods are incorporated into an application that generates test patterns for VLSI circuits. Finally, section 4.4 gives an overview how load distribution mechanisms can be used within event simulation applications using the example of time warp simulator for logic circuits.

4.1 A Middleware-Based Architecture for Load Management in Heterogeneous Distributed Systems

by Björn Schiemann

4.1.1 Introduction

In the presence of vastly growing performance needs, the coupling of several computers to multicomputer systems has become a widely adopted approach. Such a distributed system requires new management and software schemes to cope with central problems specific to its architecture. On one hand heterogeneous systems should *interoperate* irrespective of their hardware or software architecture, e.g. their operating system. On the other hand the *performance* of the entire system needs to be maximized. Interface Definition Language (IDL) technology as it is provided by CORBA-compliant IDL environments [6; 8] is one powerful solution to the problems related to the heterogeneity and interoperability of a distributed system's components. These environments are often classified as middleware because they introduce an additional communication layer located inbetween applications and the operating system. They introduce an object oriented system structure modelling a system as a set of computing and communicating entities named clients and servers. Concerning throughput and response time issues load balancing is a widely adopted solution for boosting a distributed system's performance [3; 19]. As a consequence we combine IDL and load management technology in order to exploit both their benefits. The potential of such an IDL based load balancing concept is evident:

- The overall performance of distributed systems can be significantly increased. This also holds for heterogeneous architectures.
- This concept further improves the ease of use because IDL technology introduces a better modularity. This in turn supports system management issues as well as scalability and allows for easier upgrades and modifications.

The following sections outline our architecture for IDL-based load management. The major benefits evolving from this middleware oriented approach are that the *load balancer* (LB) itself becomes an invisible object within an IDL environment. Furthermore, the load balancer is located

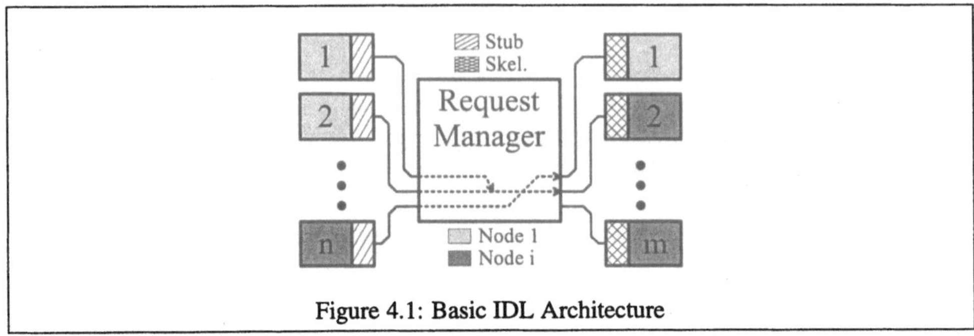

Figure 4.1: Basic IDL Architecture

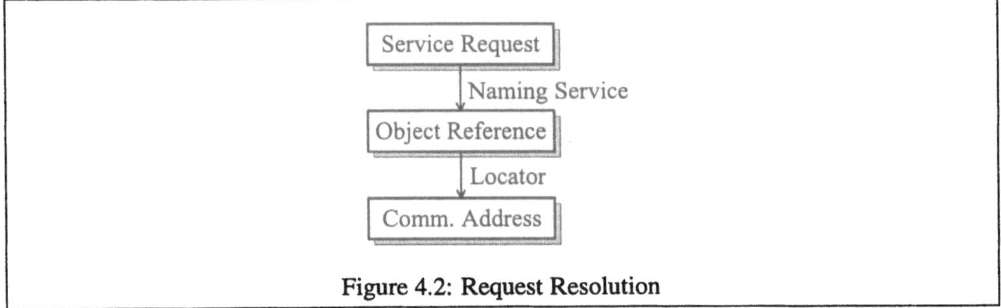

Figure 4.2: Request Resolution

close to the communication channels through which a distributed system's components interoperate. The IDL environment will be equipped with software monitors allowing to *monitor* important load information about communication patterns and data access. Furthermore, the environment provides a valuable means for *transmitting* any load information within the entire system, even between heterogeneous nodes. In conjunction with the IDL environment's capability to connect heterogeneous clients and servers this powerful and performance boosting symbiosis makes load balancing applicable even to heterogeneous distributed systems [12; 13; 14; 15]

4.1.2   IDL Technology

The heart of an IDL environment is the request manager (Figure 4.1). Its job is to manage the interaction between a system's objects: When a client requests a server to perform a service for it, the request manager provides the mechanisms for routing parameters and results between the respective objects (e.g. processes) residing on any of a system's nodes. In CORBA-compliant environments the request manager is called object request broker (ORB).

The interfaces through which clients and servers communicate are implemented by means of a dedicated specification language which is the interface definition language (IDL). An IDL compiler derives dedicated program code out of these IDL specifications. The additional code will be linked into the clients' and servers' code. On the client side the linked interfaces are called stubs whilst on the server side they are referred to as skeletons. Due to the use of these interfaces, which support an IDL specific request and data format, clients and servers may be implemented in different programming languages. They may even reside within different operating system

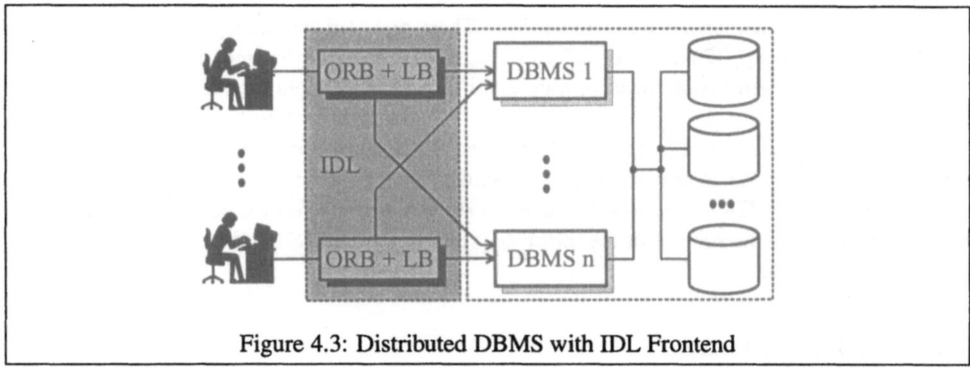

Figure 4.3: Distributed DBMS with IDL Frontend

and hardware environments. Furthermore, the ORB enables clients to call servers *transparently* without having to know where these are located. Thus clients can request services the servers provide by name (Figure 4.2). A Naming Service delivers the respective Object References. These can be seen as some IDL specific address information. Finally the locator provides the communication addresses needed for calling the requested services via remote procedure calls (RPC). The locator is an integral part of the ORB and retrieves the addresses from a dedicated database [16; 17] called implementation repository (Impl.Rep.) [9]. For optimization purposes usually proxies are used on the client side which are able to cache all address information. This helps avoiding this request resolution process for succeeding requests.

The number of IDL implementations available on the market is still growing. Many of these base on the CORBA specification [7]. Currently there are three widely used language bindings which connect to C++, C and Smalltalk. In spring 1995, the CORBA 2.0 specification was fixed [8]. Its most important enhancement is the cooperation of several ORBs via dedicated gateway mechanisms called bridges. This function is compulsory for integrating heterogeneous systems with multiple ORBs. Well-known CORBA-based products are IONA's Orbix, SunSoft's NEO [18] and IBM's System Object Model (SOM) [2]. Several IDL products base on the Distributed Computing Environment (DCE) because it provides the required basic mechanisms like RPC. DCE was developed by the Open Software Foundation.

4.1.3 Target Application

One target application of our IDL-based load balancing approach are online transaction processing (OLTP) systems, in particular database management systems (DBMS) [5; 11]. Distributed DBMSs are probably the most popular commercial applications for multicomputers used in offices and administration. They are available for different platforms including BS 2000 as well as UNIX. A very popular system is the Oracle Parallel Server (OPS).

A distributed DBMS consists of several nodes interconnected by a high speed network, e.g. an ethernet [10]. Each node is usually associated with one instance of the DBMS. The OPS is a database sharing architecture. The numerous users connect to the available nodes (Figure 4.3) and DBMS instances via another network and a DBMS frontend tool which typically is a transaction monitor (TM). Each user is associated with an own DBMS client at the moment he or she logs in to the system. These clients are associated with a fixed DBMS server until the user logs out. The load units the users generate are transactions (TA). These aim at manipulating the data contained in the system's database (DB). All the performed manipulations are persistent and therefore valid even in case of a system breakdown. A transaction is an independent atomar unit in the sense that it will always be completely executed or not at all. In systems which are equipped with a transaction monitor TAs can be represented by requests for specific transaction programs. Without such a monitor TAs would be sequences of statements given e.g. in a database query language (for instance SQL). For our IDL-based load balancing approach we focus only onto a subset of a TM's functionality which is the routing of the users' requests to the DBMS. This can be implemented by means of IDL technology.

4.1.4 Monitoring Strategy

This paragraph introduces the load balancing data that can be monitored within IDL environments if these are used for any communication within the system. Furthermore the required monitoring mechanisms are motivated.

Due to its system and programming language independent request and data format an IDL environment provides ideal mechanisms for distributing load information within a heterogeneous system. The environment can further be instrumented with monitors which give insight into the behavior and interoperability of software resources and thereby into some hardware resources as well.

The IDL-based monitors can best be realized by instrumenting the clients' and servers' IDL-defined interfaces. These are the stubs and the skeletons unless dynamic request mechanisms are used (cf. [8]). The instrumentation should be performed automatically by the IDL compiler. Furthermore, features which are specific for a given IDL environment could be exploited. For instance, IONA's Orbix provides smart proxies and filter mechanisms. If possible monitoring should be focussed only onto servers. This is due to the fact that usually a significantly larger number of clients exist which will introduce more monitoring-related overhead. If load information has to be passed from servers to clients (e.g. for correlation purposes) then CORBA's exception mechanism could be extended appropriately and quite easily because each request will return an exception by default. Normally it tells that everything performed properly.

As summarized in Figure 4.4 the following basic information can be monitored:

- Knowing the *location* and *type* of servers as well as their multiple instantiations supports dynamic (re-) assignment of clients to servers (see section 4.1.5). This is the base information for load balancing.
- The *request rate*, i.e. the number of requests issued during a given time unit, gives an estimate of which components are heavily loaded and thus might become bottlenecks. The request rate can principally be generated per server or per service, even per client.
- Another utilization related measure is the *number of clients* a server or node has. It can be interpreted as queue length.
- The *response time* of requests can also be observed. If response time is measured on the client side then it usually contains the time needed for performing any involved network accesses.
- The *size* of requests given as the number of transmitted parameters or bytes is an estimate for the impact of request handling onto the network load.

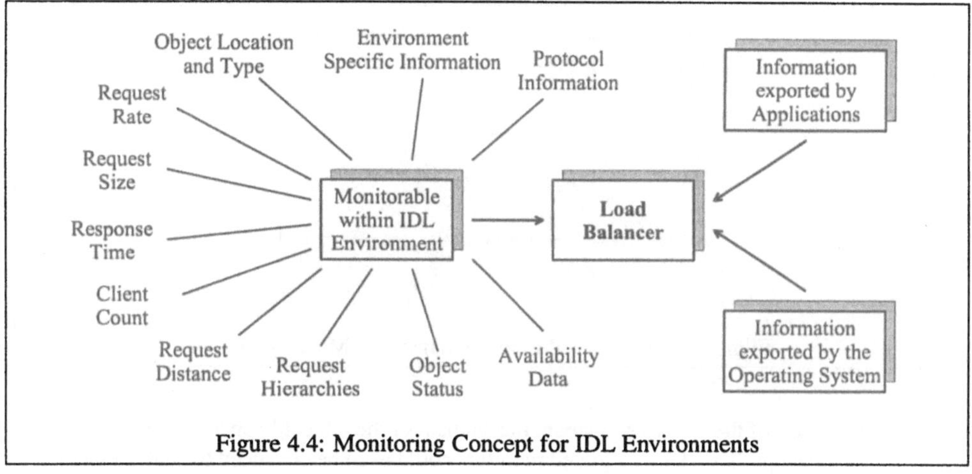

Figure 4.4: Monitoring Concept for IDL Environments

- The *request distance* refers to the number of hops a request has to perform when being routed from a client to the requested server. This is a valuable input for limiting the network traffic.

- *Request hierarchies* respectively request graphs show dependencies and cooperations of clients and servers. They describe which servers become clients again, and they are useful for supporting co-location of clients and servers.

- It might also be interesting to observe the *status* of servers and services. For instance, in some IDL environments like NEO servers will automatically be terminated if they have not been referenced for a setup amount of time. This saves resources. The load balancer can e.g. exploit this information for its server selection policy when new client-server connections have to be established.

- An IDL environment can also be used for supporting fault tolerance issues, e.g. by logging which servers or nodes are up or have suffered a failure (*availability information*). Currently this feature is not specified within CORBA and it is not supported by commercially available environments, but it can be expected to gain importance in the near future.

- Depending on the network protocol, *protocol-specific* data like request transfer time and transmission delay can be measured. One simple method for calculating the sum of transfer times needed for request and result transmission is to subtract the server's execution time from the time measured on the client side. But for better efficiency protocol-specific measures should be provided by the RPC mechanisms underlying the IDL environment.

- Corresponding to monitoring server and client specific load information it could be desirable to monitor the IDL *specific components*, e.g. the object request broker.

Many of the load parameters given above can be used as single momentary values as well as average or mean values if an appropriate history mechanism is in place which allows keeping the values' traces for a given time.

Besides of using the introduced monitoring mechanisms the load balancer will be equipped with additional interfaces through which application specific load information can be provided. Examples are resource requirements like typical memory consumption or execution time as well as some user or owner information. The mechanisms for providing this kind of information can e.g.

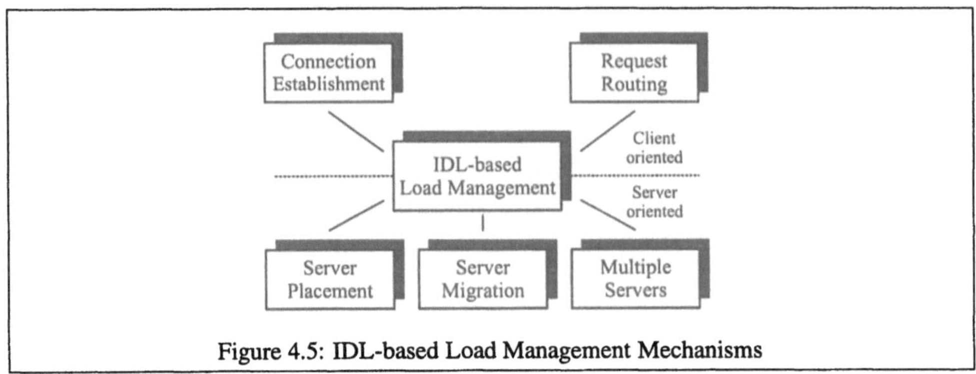

Figure 4.5: IDL-based Load Management Mechanisms

be realized by means of the CORBA attribute mechanism or the Object Properties Service [1] with its trading-like concept. The IDL-integrated monitors could even be enhanced to analyze incoming requests. Besides of being exclusively available to a single application like our distributed DBMS it is also possible that some nodes are assigned further workload. Thus there need to be further interfaces to the operating system through which the load balancer will be informed about the nodes' entire load. This holds especially when part of a node's workload does not use the IDL environment for its communication and thus is unvisible to the suggested IDL internal monitors.

4.1.5   Load Management Mechanisms

This section gives a brief overview of alternative load management mechanisms which can be realized within IDL environments. It does not select a certain static or dynamic load balancing algorithm which will use these mechanisms but leaves this decision up to the system programmer who finally implements our IDL-based load management approach. This also holds for further optimizations. As outlined in Figure 4.5 there are five basic load management mechanisms which focus on client-server connections and on server management.

Load Dependent Establishment of Connections
At the moment a client requests a server for the first time a new client-server connection has to be established. If there are several alternative servers which provide the requested service, then the load balancer should choose a low loaded server or a server residing on a low loaded node. This scheme is a load assignment strategy. It can be very successfully applied to systems with a permanent stream of newly arriving load which can be used for eliminating load imbalances. This approach is quite similar to load dependent dynamic trading [20].

Dynamic Request Routing
If there is not enough new load (permanently or temporarily) which can be used for coping with load imbalances then requests can be dynamically re-routed. This is based on breaking up selected client-server connections and re-connecting the clients to lower loaded servers or nodes. The reconnection adheres to the previously described load assignment scheme.

Creation of Multiple Server Instances
Furthermore it is also possible that the load balancer initiates the creation of additional servers if the available ones become overloaded. New and even part of the current clients can then be connected to the new server. Creating additional servers can lead to improved response times.

Placement of Servers at Creation
If new servers have to be created then it is also possible to choose the node on which a new server will be created with respect to the system's current load situation. This is very similar to the load dependent initial connection of clients to servers.

Migration of Servers
Finally the servers themselves can be moved from a high loaded node to one with a lower load. This is called migration. Migration can be preemptive or non-preemptive and usually requires the moving of all a server's data (its context).

The implementation of the first two mechanisms requires a close cooperation of the load balancer and the IDL environment's Naming Service [1] and locator. Usually the Naming Service will pick an object reference out of a number of alternatives by random. Now this will be performed with respect to the system's current load situation. If established client-server connections have to be cut then clients must be instrumented such, that they can receive the invalidation of their connections and that they will re-connect to another server. Then the load dependent server assignment scheme will be automatically applied as before. Notifying the need to re-connect to the clients can best be done either by forcing servers to reject some incoming requests and to rise an appropriate exception or by exploiting CORBA's event service [1].

The load dependent placement of servers at their creation time requires a modification of the IDL environment's server creation policy. If even temporarily deactivated servers have to be dynamically assigned to nodes when called back to life, then it will also be necessary to assign the respective server's data context to the new node. The creation of multiple server instances can be expected to require a modification of the environment's server assignment scheme. This is due to the fact that CORBA supports multiple clients per server respectively object or that a new instance will be created for each client. Thus our approach can be seen as between these alternatives. Finally it might be necessary to support the migration of servers. Load migration should only be used if no or not enough new load arrives and if severe bottlenecks arise. In general the preemptive migration of running load units causes a lot of overhead and thus only provides limited performance benefits [4]. Not only the code but also an object's data area have to be moved. In case of heterogeneous systems further problems arise if the code is not interpretable but binary because then a code translation might become necessary. The Life Cycle Service specified in [1] provides several means for migrating IDL objects, for instance the Copy() and Move() operations. Unfortunately there are a couple of unsolved problems. In particular there is no standard for updating the object references of the migrated objects which are usually maintained by the Naming Service. Furthermore, problems arise if the Object Relationship Service is used. Yet it is not possible to always correctly copy or alter the relations between objects (cf. [1]).

Only the first two of the suggested load balancing mechanisms can be applied to our target system. The main reasons behind are that our DBMS is a highly optimized proprietary product which does not rely on IDL technology and which can not be modified. It is only equipped with an IDL-based frontend (see section 4.1.3) which can be enhanced for load balancing purposes. Nevertheless the other schemes can be successfully applied to other applications.

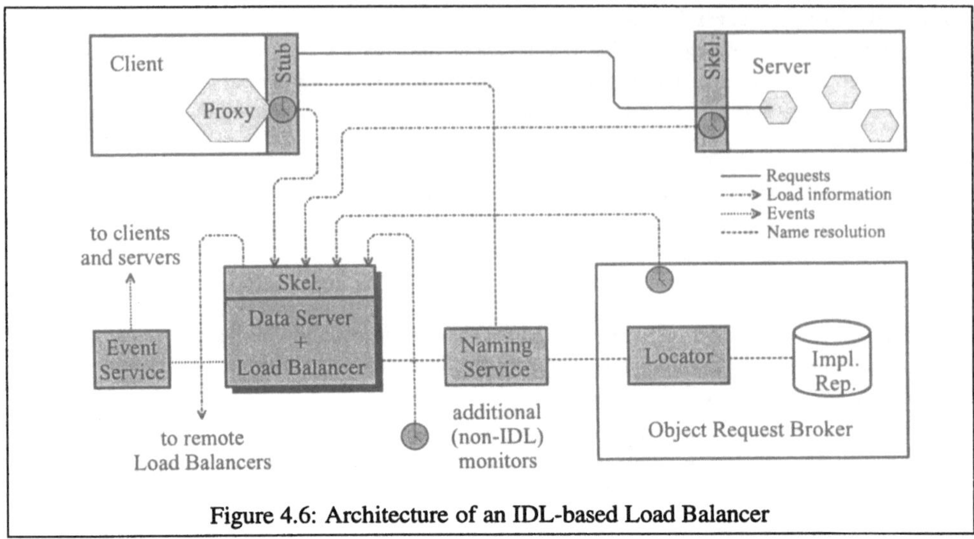

Figure 4.6: Architecture of an IDL-based Load Balancer

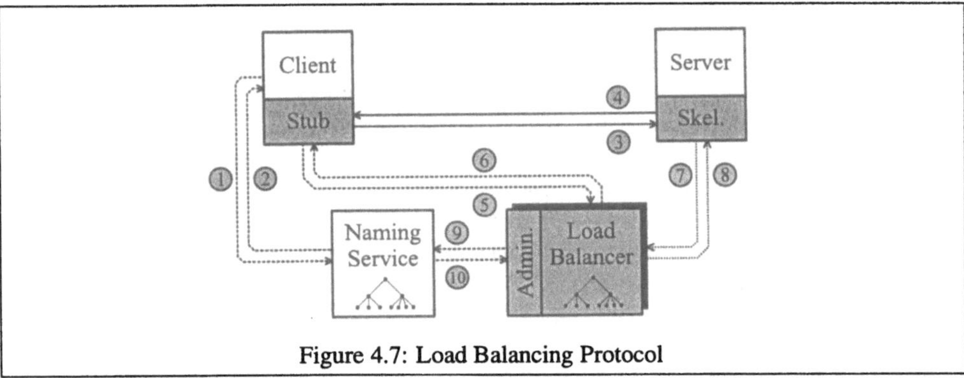

Figure 4.7: Load Balancing Protocol

4.1.6   Architecture of an IDL-based Load Balancer

As outlined in Figure 4.6 an IDL-based load balancing concept comprises a number of different components. The load balancer itself is an object in the IDL environment's object space. The load balancer is equipped with a data server frontend which collects and preprocesses the load information provided by monitors located in servers, clients or the Object Request Broker (ORB).

As outlined in section 4.1.4 additional monitors can deliver application specific or operating system specific information. The data server is further responsible for delivering load information to other load balancing instances participating in a distributed load balancing scheme. The load balancer cooperates closely with the Naming Service and the ORB's locator. For communication purposes the load balancer is equipped with dedicated IDL-defined interfaces through which load balancing instructions and load information can be transmitted. Furthermore, CORBA's Event Service could be used.

The introduced architecture of an IDL-based load balancing scheme for our target database management system application will rely on the request handling protocol outlined in Figure 4.7 . It is assumed that the ORB's locator will be called implicitly when needed. The protocol belongs to the current version of our prototype (section 4.1.7) and is dedicated to separate Naming Service and load balancer. In the future we intend to melt both these components which will make the protocol more efficient.

- Request handling without load balancing: The client asks the Naming Service for a service's object reference (1)(2) and uses it for executing the request (3)(4).
- If proxies are used on the client side which can cache the object references only (3)(4) will be used for future requests.
- Connection Establishment: A load balanced initial client-server connect will also ask the Naming Service for an object reference (1)(2). But here the load balancer will also be contacted via (5)(6) to deliver a reference which best fits the system's current load situation. Then the request will be executed as before via (3)(4).
- Dynamic Re-Connect: Via the communication mechanism (7)(8) between the server skeletons and the load balancer a server will be instructed to refuse incoming requests. Via (4) a client will be told that its previous request (3) was rejected, e.g. by throwing an exception. Thus the client stub will ask the load balancer for a new object reference via (5)(6) which will then be used for executing the request via (3)(4) as before.
- Finally the load balancer is equipped with an administration frontend which enables selective load balancing. This allows to focus load balancing on selected servers while others can be excluded. Via (9)(10) the frontend makes its own data structures coherent to those belonging to the Naming Service.

The most important *requirements* that have to be fulfilled in order to make our approach a reality are that all load balancing mechanisms and the IDL-based communication have to be transparent towards the user. Furthermore, their implementation has to be optimized in order to introduce as less overhead as possible. There must be a seamless cooperation between the load balancer and the IDL environment's components, in particular the Naming Service. Some environments perform an automatic deactivation of servers after a predefined timeout. If this is the case the Persistency Service must be used to save the load balancer's internal state during temporal deactivation. The load balancer must be capable to manage even those systems whose nodes are not exclusively dedicated to IDL-based applications which thus can be handled by the IDL environment. Even with a load balancing scheme in place the IDL environment must still be able to locate any requested services respectively their servers and to transport the requests' results back to the calling clients. Finally there must not be any negative impact onto the entire system's behavior, e.g. concerning stability and availability, and the overall system performance must be improved according to a load balancing strategy's goals, whether these aim at throughput or response time improvement.

4.1.7 Next Steps

The performance gains and benefits of our approach to combine IDL technology with load balancing will be validated by implementing a prototype for OLTP applications Figure 4.8 . The hardware comprises several Unix workstations and servers which are connected via a LAN. The

IDL environment used is SNI's SORBET 1.0 [16; 17] which is a port and improvement of Sun's NEO [18]. Its current release is compliant to CORBA 1.2 [7] and a release which is compliant to CORBA 2.0 [8] will be available soon. The DBMS application to be integrated with our load balancing enhanced IDL environment is the Oracle Parallel Server (OPS).

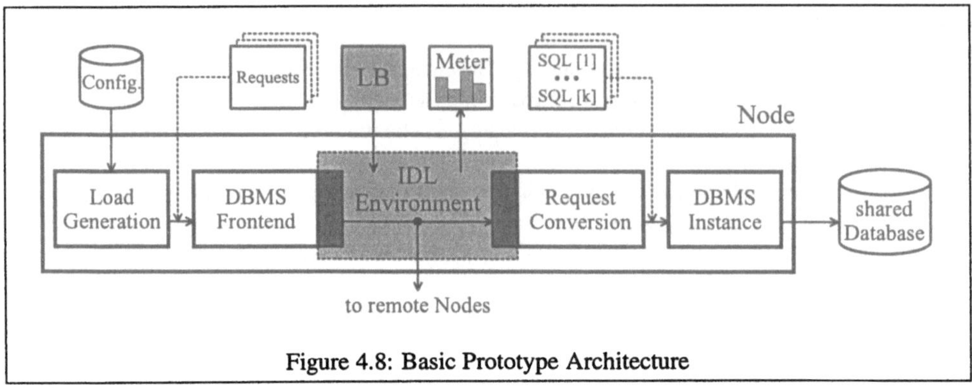

Figure 4.8: Basic Prototype Architecture

The basic architecture of the prototype is shown in Figure 4.8. It gives all components belonging to a single node, and several of these nodes will be connected and cooperate. On the very left a configurable load generator is shown. It is responsible for generating TPC B-like DBMS requests and thus emulates a predefinable number of users and terminals. TPC B is a standard benchmark. Next a DBMS frontend collects these requests and feeds them into the IDL environment which in turn delivers them to request converters. These are the connections to the DBMS instances' programming interface. For each request they call a sequence of SQL instructions from a table and fit the request's parameters into there. These sequences represent our transaction programs as they are used in systems equipped with transaction monitors. Finally the DBMS instance will execute the transaction programs.

The enhanced IDL environment provides a frame where many of today's load balancing algorithms fit in, e.g. the popular Join-the-Shortest-Queue which has proven its success in various simulations. The load balancer (LB) can redirect requests from its local node to a remote one. It is further able to dynamically re-connect clients to lower loaded servers. A meter visualizes the effect of load balancing, e.g. by showing the current throughput of a DBMS instance. It is possible to extend the meter to display even other monitors' load information. The load balancer to be realized within the prototype will rely solely onto load information that was generated by the IDL-based monitors. The implementation of the described prototype is on its way. Selected monitoring schemes are in place, and a basic load balancer has been integrated.

Our future work will include the development of a representation metric for load information which gives it an unambiguous interpretation on any of a distributed system's heterogeneous components. Furthermore we will evaluate which type of load balancing schemes is best suited for IDL-based load balancing and how the Naming Service and the load balancer could be further integrated. Finally a performance evaluation of the entire approach will be performed.

Objectives	
domain specific	no
intent	any
function	assignment
Integration Level	
monitoring	rts (middleware)
strategy	rts (middleware)
mechanisms	rts (middleware)
Structure	
problem structure	independent
subproblems	computation (CORBA requests)
workspaces	-
targets	processes (CORBA servers) and nodes

Table 4.1: General Classification

4.1.8 Classification

This section introduces a framework for middleware-based load balancing. Thus the algorithm as well as the load model and its evaluation are not fixed in advance. They can be selected according to a specific destination architecture's needs. The following tables classify the current implementation state of our demonstrator that realizes load management by assignment of CORBA requests. Further mechanisms such as migration of CORBA objects are not considered yet. Attributes of the classification that are not fixed within the framework are marked as "any".

Acknowledgements

This work was partially funded by the European Union within the basic research project ESPRIT 8144 (LYDIA).

References

[1] *CORBAservices: Common Object Services Specification*, 1995.

[2] F. CAMPAGNONI. IBM's System Object Model. In *Dr. Dobb's Special Report*, number Winter 1994/95, pages 24–28. 1994.

[3] P. DICKMAN. Effective Load Balancing in a Distributed Object-Support Operating System. In *Int. Workshop on Object Orientation in Operating Systems*, pages 147–153, Palo Alto, CA, USA, 1991.

[4] D. L. EAGER, E. D. LAZOWSKA and J. ZAHORJAN. The Limited Performance Benefits of Migrating Active Processes for Load Sharing. In *ACM Sigmetrics Conf. on Measurement and Modeling of Computer Systems*, pages 63–72, Santa Fe, New Mexico, 1988.

[5] T. HAERDER. On Selected Performance Issues of Database Systems. In *Informatik- Fachberichte 154*, pages 294–312. Springer Verlag, 1987.

[6] P. HORNIG. Die Kinder der OMA - Überblick über CORBA-Implementierungen. *OBJEKTspektrum*, (1):38–43, 1996.

[7] OMG Document, Rev.1.2, Draft 29.12.1993. *The Common Object Request Broker: Architecture and Specification.*

[8] OMG Document, Rev.2.0, July 1995. *The Common Object Request Broker: Architecture and Specification.*

System Model	
Model Flavour	none
Target Top.	het., full
Entity Top.	indep.
Transfer Model	
Transfer Space	systemwide
Transfer Policy	non-preemptive
Information Exchange	
Information Space	systemwide
Information Scope	partial
Coordination	
Decision Structure	distributed
Decision Mode	cooperative
Participation	partial
Algorithm	
Decision process	any
Initiation	any
Adaptivity	any
Cost Sensitivity	any
Stability control	any

Load Index Properties	
index dimension	multi-dim.
index type	any
aggregation time	any
aggregation space	hybrid
Load Value Composition	
index comb.	any
weight sel.	any
Model Usage Policies	
index measurement	implicit
index propagation	adaptive
model adaptivity	any

Table 4.2: Strategy and Load Model Classification

[9] OMG (Object Management Group). *The Common Object Request Broker: Architecture and Specification, Revision 2.0*, July 1995.

[10] Oracle White Paper. *Oracle Parallel Server for Pyramid Systems*, 1993.

[11] E. RAHM. Dynamic Load Balancing in Parallel Database Systems. In *EURO-PAR 96*, Lyon, France, 1996.

[12] B. SCHIEMANN and L. BORRMANN. A New Approach for Load Balancing in High Perfor- mance Decision Support Systems. In *Int. Conf. on High-Performance Com- puting and Networking, HPCN '96*, pages 571–579, Brussels, Belgium, 1996.

[13] B. SCHIEMANN. A New Approach to Load Balancing in Heterogeneous Distributed Systems. In *Proc. of the Trends in Distributed Systems Workshop*, pages 29–39, Aachen, Germany, 1996.

[14] A. SCHILL. Distributed Application Support in Heterogeneous Networks: Standards and Technological Deployments. *it+ti Informationstechnik und Technische Informatik*, 37(1):38–45, 1995.

[15] S. SHI, D. LIN and C. YANG. Dynamic Load Sharing Services with OSF DCE. In *1st Int. WS on Services in Distributed and Networked Environments*, pages 178–186, Prague, Czech Republic, 1994.

[16] Siemens Nixdorf Informationssysteme AG. *Sorbet Programming Guide; Release 1.0*, 1996.

[17] Siemens Nixdorf Informationssysteme AG. *Sorbet Tutorial; Release 1.0*, 1996.

[18] SunSoft White Paper WP 95392-001. *Solaris NEO - Operating Environment Product Overview*, 1995.

[19] M. WILLEBEEK-LEMAIR and A. REEVES. Strategies for Dynamic Load Balancing on Highly Parallel Computers. *IEEE Trans. on Parallel and Distributed Systems*, 4(9):979–993, 1993.

[20] A. WOLISZ and V. TSCHAMMER. Performance Aspects of Trading in Open Distributed Systems; Computer Communications. *Computer Communications*, 16(5):277–287, 1993.

4.2 Cooperative Parallel Automated Theorem Proving

by Andreas Wolf and Marc Fuchs

4.2.1 Introduction

Up to now, the sequential Automated Theorem Proving (ATP) has set a very powerful standard. But when dealing with hard problems ATP systems are still inferior to a skilled human mathematician. Thus, methods to increase the performance of existing ATP systems are, besides the development of new proof calculi, a focus of interest in the ATP area. One important technique to increase performance is to employ parallelism on parallel hardware or networks of workstations. Possible parallelization concepts vary from the parallel usage of different configurations of the employed provers, to the partitioning of the proof task into subtasks that are tackled in parallel. Moreover, one can expect further improvements by interactions between the different parallel provers (cooperation), e.g. by exchanging intermediate results.

In this article we want to analyze the requirements for automatic load distribution for parallel theorem proving. The article is a contribution to the discussion on the possibilities of cooperation of ATP's, especially concerning questions of load distribution. A generic approach to cooperative concepts is presented. The costs needed for implementation and runtime, especially for communication, are estimated. Furthermore, an existing parallel prover, *SPTHEO*, and a new cooperative prover based on *SPTHEO*, *CPTHEO*, are introduced.

Developers of theorem provers are often more logicians than experts on parallel hardware. Therefore, function libraries are needed for the communication between two or more processes. Tools to achieve an optimally balanced load distribution should be as transparent as possible.

We assume the existence of a widely used interface for parallel applications, e. g. the *Parallel Virtual Machine (PVM)* [11]. It makes the access to network or parallel hardware fully transparent to the user. Unfortunately, broadcasting of messages to a group of processes in *PVM* is simulated only by single sequential transmissions resulting in a decrease in performance. Another desirable function is the possibility to estimate the expected performance of the involved processors to spawn new processes to convenient machines. The following sections will consider which aspects have to be dealt within this context by an automated load distribution tool. Making migration of the processes to other processors possible can use the potential of the processors better than the performance snapshots before the spawn. A better load distribution can thus be achieved. As we shall explain later, there is, in general, an unsolvable difficulty to estimate which resources are needed for a certain prover in the context of a parallel proof system.

In the following a short introduction to parallel theorem proving is given. Section 4.2.2 contains a classification of aspects to be considered when constructing cooperative provers. Section 4.2.3 describes the *SPTHEO* prover and introduces the model of a new cooperative prover *CPTHEO* which is based on *SPTHEO*. We conclude with a classification of the load distributing mechanisms to be used by *CPTHEO* (section 4.2.4) followed by some conclusions in section 4.2.5.

4.2.1.1 What is Automated Theorem Proving (ATP)?

General Remarks. Theorem proving deals with the search of proofs of certain conjectures from a given theory. Both conjecture and theory are formulated in some language $\mathcal{L}$. An often used language is e. g. the First Order Predicate Logic (PL1). PL1 can be used to formulate many mathematical problems as well as problems taken from the real life. Unfortunately, PL1 is undecidable in general (Church, 1936). One can show that the set of theorems of a given theory is recursive enumerable. Especially, there exist proof procedures able to recognize each valid formula of a given theory after a finite amount of time. Thus, at least semi-decision algorithms can be constructed although the problem to decide the validity of a formula is undecidable in general.

In the past a lot of different logic calculi have been developed and implemented. Important properties of such calculi are the *soundness* and *completeness*. Soundness of a calculus means that it does not give incorrect answers and announce a formula as valid although it is not. Completeness means that the calculus has the potential to prove every valid formula.

Basically, proof calculi (or automated theorem proving systems based on such calculi) can be divided into two different classes. On the one hand there are synthetic calculi that work in a bottom-up manner, on the other hand one can employ analytic calculi that work top down. Analytic calculi attempt to recursively break down and transform a goal into sub-goals that can finally be proven immediately with the axioms (*goal-oriented provers*). Synthetic calculi go the other way by continuously producing logic consequences of the given theory until a fact describing the goal is deduced (*saturating provers*). For further information on these issue we refer to [3].

Note that the problem of proving the validity of a given conjecture by using a certain calculus can be interpreted as a search problem. A proof calculus is a *search calculus* that is represented by a transition system consisting of a set of allowed states, a set of rules for changing states, and a termination test for identifying final states. Basically, search calculi are divided into two classes: *Irrevocable* search calculi and *tentative* search calculi. Using irrevocable calculi, an application of a search step never needs to be undone in order to reach a final state. Tentative search calculi require provision for the case that a sequence of search steps does not reach a final state. Usually *backtracking* is needed in order to try a solution with each alternative search step that can be applied within a search state.

Independent from the search calculus to be applied, the general problem of ATP is that the search spaces one has to deal with when searching for a given conjecture are tremendous. Thus, the development of more efficient methods to reduce and examine the search space is one of the aims of the automated theorem proving community.

Model Elimination.   As a concrete proof procedure we want to mention the *Model Elimination* calculus (*ME*). This calculus can be interpreted as an analytic calculus. Furthermore, backtracking is usually needed in order to prove a theorem. In this paper we only want to recall some basic concepts. An introduction to *ME* can be found in [15]. Within this paper, tableau style representation of *ME* proofs is used. *ME* deals with clauses, so we cope with a set of universally quantified clauses as OR-connected lists of literals.

A *clausal tableau* is a tree with nodes labeled with literals. To simplify the following, it is assumed that we want to prove the validity of a literal under a given set of clauses. Thus, the initial tableau consists of the negation of the literal to be proved. Of course, more complex formulas can be treated. The tableau can be *expanded* by appending instances of the literals of a clause to a leaf of the tableau tree as new leaves. A branch of the tree is *closed* if it contains complementary literals. A substitution on the whole tree may be required to make the literals complementary.

In *ME*, after an expansion of the tree, one of the newly created leaves must be closed against its immediate ancestor. Additional closing of other branches after an expansion is called *reduction*. Without loss of generality we can assume that one of the nodes closing a branch is a leaf. A tableau is a *proof* if all its branches are closed.

In order to prove a given literal it is necessary to systematically construct all possible tableaux. Since usually a lot of different expansion and reduction steps are applicable to a given tableau one has to search for a proof in a *search tree* whose nodes are marked with (different) tableaux. A node in this search tree that represents an open tableau is also called a proof task.

4.2.1.2 Developing Parallel Provers

Fundamentals.   There are powerful sequential automated theorem provers in the theorem prover community. As examples we give here the (incomplete) list *DISCOUNT* [2], *OTTER* [17] and *SETHEO* [13] which belong to the most popular automated theorem provers. But when dealing with "hard problems" these provers are inferior to a skilled human mathematician due to the tremendous search spaces the provers have to deal with.

The performance of sequential provers, however, can often be significantly increased by developing parallel versions of them. The following table classifies examples of existing parallel provers according to the criteria given below.

	non-cooperative	cooperative
partitioning completeness based	PARTHEO [14] METEOR [1] SPTHEO [26]	PARROT [12] ROO [16] DARES [4]
partitioning soundness based	MGTP/G [10] SPTHEO [26]	
competitive different calculi		HDPS [23]
competitive unique calculus	RCTHEO [18] SICOTHEO [20]	DISCOUNT [2]

Parallel theorem provers can be *cooperative* or *non-cooperative*, depending on their behavior in information exchange. *Cooperative* systems exchange information during a proof run and not only in the phases of their initialization and termination. The paper especially deals with the possibilities to design provers of this kind (cf. the following sections).

Theorem provers can *partition* their work, for instance splitting the whole task into a set of subtasks, or partitioning the considered search space. Partitioning systems can be classified analogously to the usual classification of parallel algorithms into *OR and AND parallel* algorithms: *Completeness based* systems create completely independent subtasks when partitioning the search space, i. e. the solution of one single problem means the solution of the whole problem. *Soundness based* systems require the combination of the solutions of the subtasks to a common solution. These two strategies can be combined, so that sub-provers can deal with independent and dependent subproblems. Partitioning systems usually use instances of the *same inference machine*, because the partitioning of the search space is difficult for different logic calculi or even for different variants of the same calculus.

Competitive systems are proof systems where each system works on the whole proof task. But the systems differ in the way how the involved provers are parameterized. Even the use of different logics in parallel is possible. In constructing competitive systems, it seems to be useful to encourage *different calculi*. The advantages of each of them can balance disadvantages of others. This strategy assumes that for each of the provers involved a class of tasks exists where it performs better than all other provers included in the common system.

Developing parallel provers is sensible since the parallelization of a proof system yields the potential to achieve speedups compared with the original system. Partitioning the search space offers the possibility to reduce the amount of time spent with exploring unnecessary parts of the search space. This is due to the increased chance that at least one of the involved provers immediately starts exploring an "interesting" part of the search space (assuming that the search space is completely assigned to the different provers). Competitive systems can result in super-linear speedups because one has a higher probability to use a search strategy that is well suited for a given problem. Since usually a lot of different configurations of a prover are imaginable and no a priori knowledge is available to decide which configuration is well suited for a specific problem it is reasonable to use different configurations in parallel.

The highest gains in efficiency, however, can be expected when using cooperative systems due to the expected synergetic effects. Thus, the development of cooperative systems is a main research area when dealing with parallel theorem provers.

It is known that on relatively easy problems parallel provers require a higher amount of time than a similar sequential prover due to the enlarged overhead the parallel system needs to launch the program. Furthermore, communication overhead occurs that additionally decreases the inference rate of the involved provers. But this disadvantages will be compensated dealing with really hard problems if synergetic effects occur.

So it should be the aim of parallel provers to obtain profits in those domains where existing sequential or non-cooperative parallel systems do not find any proof, or even if they find one, only with a comparatively long amount of time. During the development of the parallel version of *SETHEO* dealing with partitioning of the search space (*SPTHEO* [26]),a comparison with the results of the sequential version was performed using a runtime of 1000 seconds. Problems of this complexity seem to be convenient examples to test the performance of parallel provers.

4.2.1.3   Cooperation – reasons and a definition

Sequential search procedures often employ a large set of heuristics and refinements of the underlying calculus to prevent unnecessary search, e. g. to avoid redundant search steps. If the search is shared among different processes, it may be useful to exchange information on such redundancies, as well as on important intermediate results. Basically, there are two methods to exchange data between theorem provers: *demand driven* (as realized in the *DARES* system) and *success driven* (as realized in *DISCOUNT*). Demand driven cooperation means that a prover is interested in certain information and asks the other provers for it. Success driven cooperation means that a prover communicates information it judges to be important to the other provers. As we can see in the table below, the unlimited exchange of information usually is not sensible due to the tremendous number of information that is generated during the search.

The table shows two examples from the *TPTP* [24]. It contains the number of proof tasks the original proof task can be simplified to after *level* many inferences of a ME based prover. Thus, it also shows the number of possibly generated messages (when asking for solved tasks of another prover). One observes the importance of restricting and filtering information to be exchanged.

example	level	original	sort|uniq (percent)	subsumption (percent)
BOO003-1	7	272	83	72
	8	1006	61	36
PRV007-1	5	145	84	79
	6	826	70	55
	7	5677	65	51

In this article we want to employ the following definition of cooperation:

Definition. An automated theorem prover is cooperative, if and only if it offers on request immediately all his data and results to other provers. Immediately means that the prover utilizes the available bandwidth of the communication channels and handles the request with a higher priority than its own proof attempt. ◇

4.2.2 Cooperative Theorem Proving

In the sequel, we discuss a classification of the problems that have to be solved when constructing a cooperative parallel theorem prover with some remarks on communication and load distribution needs. The aspects to be considered can be divided into two parts.

- *How* can cooperating provers work together, i. e. which topics concerning the system architecture have to be considered?
- *What* can they do to work together, i. e. which kinds of cooperation can occur?

4.2.2.1 System Architecture

The problems to be considered in relation to the *how* part, i. e. the system architecture, can be classified as follows.

- Are the involved inference machines of the same type? Are different types of inference machines used, i. e. is the parallel system *homogeneous* or *heterogeneous*?
- Is the exchange of information planned in that way that messages must be confirmed by the receiver or not? Shall the processes wait for some events or not? These questions lead to the decision on a *synchronous* or *asynchronous* mode of information exchange.
- Which scheme of process control and control on the progress of the proof shall be selected, i. e. which *hierarchical structures* of sub-provers occur?
- Proof procedures can be saturating (deduce all valid formulas until the proof tasks occurs) or goal oriented (reduce the proof task until all subtasks can be solved using known - given - formulas). It is possible to construct a system with only *goal oriented*, only *saturating*, or with *hybrid* proof processes.

Homogeneous and Heterogeneous Systems. Parallel provers developed up to now ordinarily use (with minor changes) the same implementation of the inference machine as in their sequential incarnation. These sub-provers are united by partitioning the search tree in a completeness or soundness based manner as previously described, or they can use different search strategies in their different instances.

Due to the usage of the same inference machine and an equal coding of formulas and control parameters, the expected additional costs for the implementation of an information exchange in such *homogeneous systems* are relatively low. Moreover, we can assume that heuristics estimating the needed resources for a proof task (a process, in terms of load distribution) can be found much easier than in the heterogeneous case.

Homogeneous systems do not necessarily require the same search procedure for all involved provers. For example, the connection of *SETHEO* [13] with the *DELTA Iterator* [19], both based on the same inference machine, combines the *top down* search with the *bottom up* one. The top down prover can integrate lemmata of the bottom up prover in its proof tree. That kind of cooperation was tested successfully in the sequential case. Homogeneous systems allow an easy way to partition the search space of the involved systems by control parameters for the provers and startup configurations containing different pre-calculated proof segments.

So we can classify homogeneous systems into

- systems using the *same* inference machine with the *same* search procedure and
- systems using the *same* inference machine with *different* search procedures.

Heterogeneous systems probably need more efforts for implementation because of the necessary syntactical transformations and different semantics. In the interactive proof system *ILF* [5] the main part of implementation required for communication is syntactical transformation and adaption of theories. Considering the runtime information exchange, the context needed for different systems will significantly increase the amount of information to be exchanged. Different instances of the same inference system canonically interpret received formulas in the right way. Different systems obviously need additional information on the used calculus and the coding of proof structures within the program.

Sometimes a problem can be separated into two relatively independent subproblems which belong to different problem classes. In such cases it makes sense to give these subproblems to different provers with the best adaption to the problem considered. This is not necessarily an argument for heterogeneous structures, but absolutely an argument for the usage of special provers as subsystems.

Heterogeneous systems are justified only if the connection of the systems leads to a significant increase of the performance of the common system. That can be done for instance by integrating a prover for equational problems as *DISCOUNT* [2] into a system that is only poor on equations. Another possibility is the integration of provers applying meta-mathematical knowledge as for instance *TreeLat* [5], a special prover and model checker for lattice ordered groups.

A similar approach is the integration of model checking for instance by using *FINDER* [21] as a semantical tool into a common proof system which will reject many intermediate proof tasks to be false. If model checking is not integrated into the inference machine itself, the model checker has to be invoked such that the communication connection to the prover can reach a high throughput of data.

Heterogeneous systems can be classified into

- systems using *different* inference machines with the *same* communication language,
- systems using *different* inference machines with *different* communication languages,
- systems using not only inference machines but also *meta-methods* for the proof.

Synchronous and Asynchronous Exchange of Information. Cooperation between provers needs exchange of information at runtime, not only during a short phase of initialization and termination of one or more of the processes. Communication will take place during the whole time the proof system works. In principle, the information exchange can be done according to one of two general models: information can be exchanged *synchronously* or *asynchronously*.

Using the *synchronous* mode, all partners are informed on the situation of all other partners. A message generally will be confirmed, that means, at least one of the exchanging partners is waiting until the others are ready to receive or to send a message. The synchronous mode of information exchange has the advantage that all involved processes mutually know their standard of information, i. e. if a message was written, it is read at the same predetermined time in the program scheme.

Communicating *asynchronously* the sender cannot assume that the receiving prover got the message at the predetermined moment in the program cycle. But assuming the *message passing concept* for instance of the *PVM* [11], it is guaranteed at least that the temporal sequence of messages between each two of the involved processes will be preserved. Furthermore, it is guaranteed that all messages reach the receiver, if this process is still existing. This method has the advantage that no time is wasted by waiting for a communication partner. It has the disadvantage that usually an information is not available at the moment it would be needed. However, an important information (for instance a lemma) may not be available in time. The following cases are imaginable.

1. The information read is already available in the prover. That means that this information is redundant (at this moment). Thus, the effort to find that information is done twice.
2. It may happen that the lemma read is more general than a subgoal proved in the prover internally. Then it is possible that later subgoals can be solved using the lemma but not with the internal subproof. In that case the internal subproof should be replaced by an application of the lemma.
3. The lemma may be a contradiction to assumptions of the internal proof. Then the subproofs which employ such an assumption occurs, must be corrected.

It would be ideal, if the lemma solves an actual subgoal or needs only a few inferences to solve this goal.

Asynchronous event-oriented control can be implemented using the *non-blocking* read routines of the *PVM*. Another way is to use signals of the UNIX operating system.

Hierarchical structures of sub-provers. The structure of the processes belonging to the parallel prover can be selected according to the following models.

- All sub-provers are *on the same hierarchical level*. The start of the whole proof system is very easy in this model, because the processes do not need to take their place in a hierarchical order during the initial phase. Thus, no communication is needed for that purpose. But using this model, it is not possible to connect exactly two particular processes. Either broadcasting has to be performed resulting in high costs or the processes must get knowledge about the way they can communicate, for instance using process tables.

- The sub-provers can be arranged in a *hierarchical structure*. If a prover generates new subtasks, it will create the subordinated provers and transmit the tasks to them. The information exchange can be done easily in this model but the controlling of the globally used resources is complex. The uniform distribution of the generated processes according to the processor load to the involved machines would have to be realized using *PVM* or additional software.

- The *combination* of different structures following both models is possible, too. Nevertheless, each architecture needs facilities for information exchange, i. e. it should be possible to group processes and perform broadcasts to such groups. Information on the structure of the whole prover, e. g. its hierarchical organization and the information exchange relations might be a hint for effective load balancing, too, and so it should be accessible to a load distribution system.

Combination of Goal Oriented and Saturating Provers.   It is the aim of using parallelization of theorem provers to decompose the search space for finding a proof, and to treat the parts of that search space at the same time in parallel. Using cooperative concepts it is possible to exchange information about solved subgoals to prevent redundancies, especially to solve a subgoal more than one time. Furthermore it is possible to partition the search space in a "horizontal" manner, i. e. to combine *bottom up* with *top down* proof procedures.

This combination already was successfully tested in a version of the theorem prover *SICOTHEO* [20]. In that system the *DELTA iterator* [19] and *SETHEO* [13] worked together sequentially in the sub-provers. The depth of the found proof (i.e. depth of the corresponding closed tableau) usually is much lower compared to a single run of *SETHEO*.

In an application of the interactive *ILF system* [5] a combination of *DELTA* and *SETHEO* was used (only one process of each kind, both working sequentially). In the domain of lattice ordered groups (with a CPU time resource from 30 up to 120 seconds) it was possible to increase the average depth of the proofs including the (later) expanded lemmata from 5 to 8, compared with the standard version of *SETHEO*.

Another existing application is the combination of saturating and goal oriented *experts* in *DISCOUNT* [6] which works on equational problems.

Using a sequential prover, combining top down and bottom up mostly means that at first some steps bottom up are performed followed by top down calculations. So in parallel a high priority to the bottom up processes should be assigned in the beginning, decreasing while the prover works.

4.2.2.2 Kinds of Cooperation

In the above part, the discussion was about aspects of "how" provers can cooperate, not strongly depending on the special inference machines used in the parallel prover. In the following, we will look "into" the sub-provers, and we discuss "what" the involved inference machines can do together. A classification of kinds of cooperation is given as follows:

- the exchange and optimization of configurations and control information,

- the exchange of intermediate results (lemmata), and
- the exchange of failure information.

Different Strategies of Search.   In most cases, the provers involved in a parallel proof system partition the search space. For example *SPTHEO* [26] expands the search tree up to a certain depth, and the resulting tableaux will be given to the involved single provers. These provers "replay" their initial tableau and start their search on that base. The further search of the sub-provers was done using the same strategy for all provers. This technique yields very good results [26].

Unfortunately, it is possible that the same search strategy produces similar subgoals at the same time on similar proof tasks. That means loss of chances that intermediate results solved by one prover could be interesting for another one: the other prover has already solved that subgoal itself and cannot use the external results.

If different search strategies are used by the sub-provers (different kinds of *iterative deepening*, most possible *length* and *linearity* of inference chains, search for goals of a certain *structure*, etc.) it is possible that intermediate results generated by one prover can be re-used by one or more of the others for their further work. Thereby, it can be accepted that some of these provers work with *incomplete search strategies*, if they solve tasks from their specific domain especially fast and efficiently. But it should be realized that the whole system remains *fair* and *complete*. So sub-provers with specialized strategies can be used for instance to generate lemmata or to deal with subgoals belonging to special problem classes such as equality problems.

The cooperation of parallel provers can be classified with respect to used search strategies:

- provers with the *same* search strategy,
- provers with *different* search strategies, *each* fair and complete,
- provers with *different* search strategies where only the *ensemble* is complete, and,
- provers with *different*, even in the ensemble *incomplete* strategies.

Cooperation and Competition.   The relation between cooperation and competition has been discussed in [9] and [25]. Often, competition is the basic concept of existing parallel provers. The reasons are the low cost of implementation and the small amount of interprocess communication.

Cooperation and Competition do not need to be a contradiction. If competitive prover exchange information, they lose time for their own work, but using the results of their competitors they may solve their tasks faster. An example for such a synergetic effect is the *Teamwork Method* [8]. That concept includes the competition of some provers which exchange their intermediate results periodically.

Cooperation *without* competition inheres the risk that sub-provers with bad results on a special class of tasks can decrease the performance of the whole system, if their useless results increases the amount of information to be processed. Competitive concepts can eliminate such provers from the actual configuration of the system.

Exchange and Optimization of Configurations and Control Information.   Speaking about cooperation in the context of parallel theorem provers, one at first thinks about exchange of formulas or sets of formulas. But, a further possibility is the exchange of all that information that describes *how* an inference machine works such as

- control parameters,
- used heuristics,
- search strategies,
- inference rates, and so on.

It is useful to terminate the sub-provers that were less successful in the last time period and to start new provers with the control information of those of the more successful ones. It is also imaginable that the existing provers with lower success get the parameters to change their behavior in the intended sense.

Based on this idea a system could be implemented where a *SETHEO* based parallel prover optimizes a set of control parameters. During a certain interval of time some span of parameters will be tested and only the successful settings will be kept in the next interval where the span can become smaller or the kind of the parameter can be changed.

Furthermore it is to realize that in this model information is exchanged only infrequently. An information transfer only after some cycle of stand alone work is used less frequently than a continuous transfer during the proof process.

It has to be considered which criteria are suitable to determine the quality of the control configuration of a prover. A measure for that purpose can be for instance the depth of the proof structure relative to the other sub-provers or the number and quality of the generated lemmata. Using such criteria it should be considered that the resulting system must have a *fair* search strategy to save the completeness of the whole joined system.

Considering existing systems, results of *DISCOUNT* [7] show that generating proof procedures improve searching by means of exchange. Analytical provers probably will cause more problems.

Exchange of Intermediate Results (Lemmata). In the previous paragraph, we described the exchange of configurations. Now, we discuss the concept one thinks about at first, namely the exchange of proved intermediate results (proved formulas). Considering their structure, the formulas to be integrated will mostly be *literals*. But it is also possible that *more complex formulas* have to be transmitted.

Considering the way how these lemmata are exchanged it should be possible to integrate new axioms into the knowledge base during the inference process. It should be easy done using literals (facts) but the possibility to integrate more complex formulas, especially clauses, would be desirable.

If that is not possible, the old situation of the inference machine has to be conserved and must be reconstructed after the integration of new formulas into the knowledge base. A possible technology to do that is the *description of start configurations* distributing tasks, as it is used in *SPTHEO* [26]. That method probably has lower costs of implementation than the variant of integration into the running prover, but it has other disadvantages, for instance the transfer of context that is useless for the generated subgoal or even harmful in the new context. Altogether we get

- provers integrating lemmata "on the fly" without restart, and
- provers restarting after lemma integration.

We can assume that during the proof process a large number of lemmata can be generated, as each inference step can lead to a new lemma. This will be the main conceptual problem on the way towards a cooperative parallel prover. A transmission of all possible lemmata without filtering, even to a subset of the involved sub-provers, will cause an overloading of the network. Thus, the candidates for being a lemma must be evaluated to decide if they can be exchanged. In order to get a *measure of a lemma* one can use information on

- the syntactical structure of the lemmata (e.g. the high generality), and
- information on the derivation of the lemmata, e.g. the number of inferences.

If provers can ask for goals that are important for their work, then the existence of such a question for a formula should be a criterion to be a lemma. These questions should be filtered analogously to the lemmata to avoid network overload. The concept of proof requests is implemented in *DARES* [4] where information was exchanged if a formula has already been proved by another sub-prover.

Exchange of Failure Information. In the sections before, we discussed the exchange of information about successful events (successful configurations, lemmata). These informations can be considered to be *positive* knowledge. Furthermore, *negative* knowledge can be exchanged, for instance, the provers can communicate about, what they can*not* prove. Such messages should include the conditions under which a proof failed, i. e. the parameter settings and search bounds.

This concept also demands a strong selection, as it is already explained considering lemma and request generation. It must be realized that not every *backtracking* step generates negative information. It should be possible to use analogous criteria as in case of lemma generation.

4.2.3 SPTHEO and CPTHEO - two applications

As a basic approach we consider the sequential Model Elimination [22] style Theorem Prover *SETHEO* which was developed at the Technische Universität München. It is the basic inference machine involved in the prover systems described in this section. Problems in constructing parallel provers with other inference machines should be similar to our consideration, at least with respect to the generic needs of communication and load distribution.

4.2.3.1 The parallel theorem prover SPTHEO

Static Partitioning with Slackness (SPS) [26] is a method for parallelizing search-based systems. Traditional partitioning approaches for parallel search rely on a continuous distribution of search alternatives among processors ("dynamic partitioning"). The SPS-model instead proposes to start with a sequential search phase, in which tasks for parallel processing are generated. These tasks are then distributed and executed in parallel. No partitioning occurs during the parallel execution phase. The potentially arising load imbalance can be controlled by an excess number of tasks (slackness) as well as appropriate task generation. The SPS-model has several advantages over dynamic partitioning schemes. The most important advantage is that the amount of communication is strictly bounded and minimal. This results in the smallest possible dependence on communication latency, and makes efficient execution even on large workstation networks feasible. Furthermore, the availability of all tasks prior to their distribution allows optimization of the task set which is not possible otherwise.

SPTHEO is a parallelization of the *SETHEO* system, based on the Static Partitioning with Slackness (SPS) model for parallelization. It consists of three phases.

- In a first phase, an initial area of the search space is explored and tasks are generated. The number of generated tasks exceeds the number of processors by a certain factor (slackness).
- In a second phase, the tasks are distributed.
- Finally, in a third phase the tasks are executed at the individual processors.

In this model the search space is initially developed sequentially until a sufficient number of alternative sub-search spaces (tasks) have been generated. These tasks are distributed to multiple processors that search the alternatives in parallel until one finds a proof. The number of tasks generated typically exceeds the number of processors, and the extent of this is the "slackness" in the system. For reasons of practical search completeness, each processor executes all its tasks concurrently (preemptive execution). *SPTHEO* is implemented in C and PVM, and runs on a network of 110 HP workstations. Extensive evaluations showed significant performance improvements over *SETHEO* and a previous parallelization of it.

Figure 4.9 displays the number of problems from the TPTP [24] solved by SETHEO within runtime limits ranging from 0 to 1000 seconds and ($\approx$ 17 minutes).

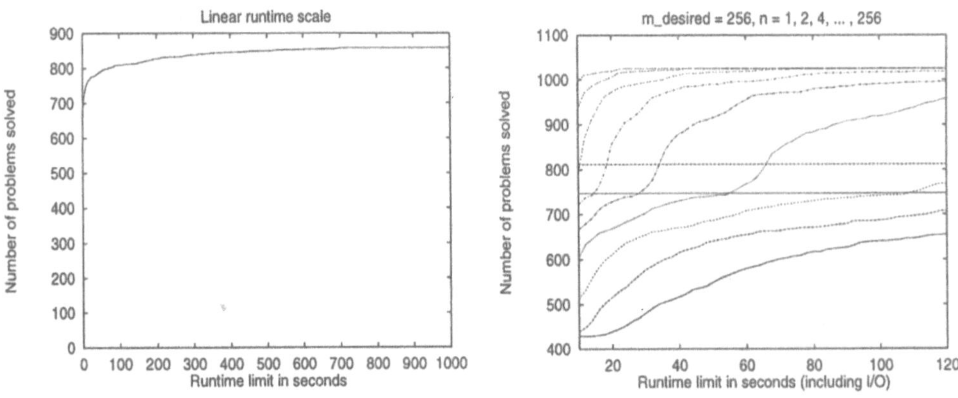

Figure 4.9: Left: The number of problems solved by SETHEO as a function of the runtime limit. Right: The performance of SPTHEO for overall runtime limits between 10 and 120 seconds (20 seconds per task). Each curve denotes a particular number of processors, starting with one processor for the lowest curve and representing twice as many processors with the next higher curve. The horizontal lines show the SETHEO performance for runtime limits of 10/120 seconds.

As an example, given a runtime limit of 1000 seconds, SETHEO solves 858 problems. It can be expected that increasing the runtime limit only leads to a small increase in the number of additionally solved problems.

An examination of the proof-finding performance of SPTHEO for different runtime limits is also given in figure 4.9. It shows the number of problems solved by SPTHEO for 256 processors. The plot is for 256 generated sub-tasks and 1 to 256 processors. Each curve shows the performance for twice as many processors as for the lower one. The asymptotic upper bound on the number of proved problems is due to the runtime limit of 20 seconds per task.

4.2.3.2 A model of the cooperative prover CPTHEO

In this section the most significant components of the *CPTHEO* model are explained:

- the cascading task generation of the top down inference machine,
- the unit preferring lemma mechanism,
- the redundancy filtering methods, and
- the relevance measuring mechanism.

Concluding, a scheme of the prover is given.

The cooperative prover *CPTHEO* to be developed uses techniques of *SPTHEO* for the scheduling of tasks and the success control. In addition to *SPTHEO*, the partial evaluation of the search space to determine the tasks of the sub-provers will be done iterative. After each iteration (and after filtering) the generated tasks will be labeled with a number. That number measures the expected importance for the further proof process. The label mechanism is a tool to control the directions of the proof trial, and it can be used for effective load balancing. Due to the iterative development of the set of proof tasks, an adaption of the whole proof attempt to the real hardware configuration is possible even during the proof process. Furthermore, by changing the labels of the proof tasks it is possible to influence the heuristic of the search at runtime.

The information exchanged between the involved sub-provers consists in its main part of unit lemmata and unit proof requests (failed proof attempts of top down provers). Units are preferred because of their lower complexity as formulas and their higher expected re-usability for other sub-provers. Furthermore, the *SETHEO* inference machine is constructed to read only units at runtime. Experiments showed that the unit preferring heuristic is only a weak restriction, because it is quite probable that units occur.

The filters of the sub-task generators and the lemma and proof request generators consider the following:

- identical formulas, subsumed formulas,
- tests with models of the considered theory,
- variants (in experiments very powerful),
- models given of the human,
- model fragments (if only infinite models exist).

Filters only delete multiply occurring or obsolete formulas. Referees rank the sub-tasks of a proof as well as lemmata or proof requests by labeling them with measures. Using these measures the proof process is controlled. The measures depend on:

- the generality of a literal,
- the derivation cost of a literal (the number of inferences to deduce the literal),
- the relatively isolated position of a subproof leading to a literal that is for instance only a few or no reductions to literals outside that subproof considering *SETHEO* proofs,
- the multiple usability of a literal in the considered proof or in the already existing parts of the proof, and
- the similarity of the generated facts to the task to be proved.

The scheduling has to guarantee the fairness of the whole proof attempt, and so the completeness of the proof procedure.

Now we describe the components. *SETHEO* provers reduce tasks from the task pool and generate new ones as well as proof requests. The *DELTA* iterators generate lemmata. The filters delete redundant tasks and lemmata from the task pool. The referees rank proof tasks, proof requests and lemmata due to their expected importance. All these processes are controlled by a central supervisor that starts and finishes processes and guarantees the liveliness of the whole prover system. To avoid bottlenecks in the information transmission, the task pool is kept locally by the involved processors. Only a certain amount of tasks labeled with high ranking measures is sent to a central task pool. The rest can be sent on request.

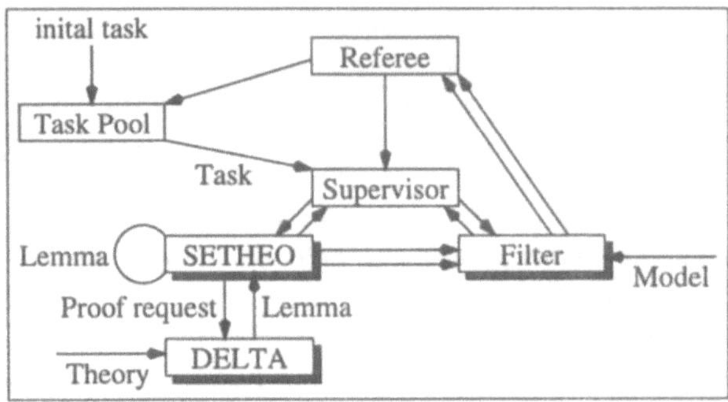

4.2.4 Classification

In the following section, the concept of the cooperative prover *CPTHEO* is classified according to the classification schemes commonly used within this book.

The load distribution mechanism of the prover is domain specific. It will use methods for load distribution on tasks representing partitions of a search based problem. The resources needed to solve a specific task can mostly not be pre-determined. The cooperative prover has to share resources with other applications. Resource usage of other applications cannot be explicitly controlled. Therefore, the parallel application has to adapt its workload distribution also to the actual resource usage of other applications. The prover generates new sub-tasks at runtime which have to be assigned to certain targets. Considering the integration level, the monitoring of the relevant information for load distribution is done within the prover. Functions to realize the general strategy come from underlying systems as they are described in this book and are not a part of the prover itself. The load distribution in *CPTHEO* is based on placement of subproblems, i. e. the problem structure is location independent. The generated sub-tasks are assigned to processes.

Consider the load distribution strategy of the cooperative prover. The model flavor is fairness. The fair use of resources motivates the distribution. First of all, the prover is designed to work on clusters of workstations. It is planned to adapt it to a massive parallel hardware. New subtasks are placed systemwide due to the transparent view at the workstation cluster. Processes, once placed, stay on their attached processors until termination. So the model is non-preemptive. Information on their actual amount of work can be exchanged between any two sub-provers. The load distribution mechanism can collect all data available to solve its distribution task. The decision on new tasks to create is organized hierarchically in *CPTHEO*. All involved provers take part in the

I. General classification

Objectives	
domain specific	yes
intent	adaptive
function	assignment
Integration Level	
monitoring	application
strategy	runtime
mechanisms	runtime
compiler support	none
Structure	
problem structure	independent
subproblems	processes
workspaces	objects
targets	processes

II. Strategy classification

System Model	
Model Flavor	fairness
Target Topology	NOW
Entity Topology	independent
Transfer Model	
Transfer Space	systemwide
Transfer Policy	non-preemptive
Information Exchange	
Information Space	systemwide
Information Scope	complete
Coordination	
Decision Structure	hierarchical
Decision Mode	cooperative
Participation	partial
Algorithm	
Decision process	dynamic
Initiation	central
Adaptivity	fixed
Cost Sensitivity	high
Stability control	partial

III. Load Model

Load Index Properties	
index dimension	single-dimensional
index type	application defined
aggregation time	comprehensive
aggregation space	system related
Load Value Composition	
index combination	none
weight selection	not applicable
Model Usage Policies	
index measurement	periodically fixed
index propagation	after measurement
model adaptivity	fixed

Table 4.3: General, Strategy and Load Model Classification

decisions on the placement of new processes by their relevant data. The load distribution within the cooperative prover is a dynamic one depending on the actual load at runtime. Communication to access load data is initiated by a central control unit. The strategy of the algorithm is fixed. The costs of load distribution are intended to be as low as possible to achieve the highest possible performance of the prover with only a minimal overhead. The stability of the whole system is partially guaranteed by the process scheduling mechanism, and also by the proof procedure itself.

The load index is determined by the PVM and its special add-ons. It is derived from only one parameter which comes from the underlying hard- or firmware. The load index is aggregated for a certain period of time. The index measurement is invoked periodically after a fixed amount of time, the load computation model of the prover is fixed, and the results are propagated to the decision processes after the measurement.

4.2.5 Conclusions

The increased power of theorem provers tends to fit automated theorem provers to solve more than only toy examples in the near future. Today ATP systems support the human for example in interactive proof environments as *ILF* [5]. Within *ILF* provers fill in steps of the proof sketch the human writes down. The more the performance of the automated provers increases, the larger the gaps between the given proof steps can be. One possibility to increase the performance is the parallelization of provers. Especially cooperative systems have the need for fast communication between the involved processes and for a transparent load balancing due to the unpredictable progress of proof attempts.

It is known that on relatively easy problems parallel provers have a higher amount of time than a similar sequential prover due to the enlarged overhead the parallel system needs to launch the program and initial communications. So it can be expected that parallel provers, dealing with more communications at run time in addition, have once more lower rating than the corresponding sequential systems, if they have to deal with relatively easy problems. But this disadvantage will be compensated dealing with really hard problems, if synergy takes effect.

So it should be the aim of the development of a new cooperative parallel prover to obtain profits in these domains where existing sequential or non-cooperative parallel systems do not find any proof, or even if they find one, they do that only with a comparatively long amount of time.

Up to now, *CPTHEO* exists as a model as described in this article. Prototypical implementations of details of the prover showed the potential of the cooperative concept. An implementation of the whole system is planned for the near future.

References

[1] O. L. ASTRACHAN. *Investigations in Model Elimination based Theorem Proving*. PhD thesis. Duke University, USA, 1992.

[2] J. AVENHAUS, J. DENZINGER and M. FUCHS. DISCOUNT: A System For Distributed Equational Deduction. In *Proc. 6. RTA*. Springer, 1995.

[3] W. BIBEL. *Automated Theorem Proving*. Vieweg, 1982.

[4] S. E. CONRY, D. J. MACINTOSH and R. A. MEYER. DARES: A Distributed Automated Reasoning System. In *Proc. AAAI-90*, 1990.

[5] B. I. DAHN, J. GEHNE, T. HONIGMANN, L. WALTHER and A. WOLF. *Integrating Logical Functions with ILF*. Preprint, Humboldt University Berlin, Department of Mathematics, 1994.

[6] J. DENZINGER and M. FUCHS. Goal Oriented Equational Theorem Proving. In *Proc. KI-94*. Springer, 1994.

[7] J. DENZINGER and M. KRONENBURG. *Planning for Distributed Theorem Proving: The Team Work Approach*. SEKI-Report SR-94-09, University of Kaiserslautern, 1994.

[8] J. DENZINGER. Knowledge-Based Distributed Search Using Teamwork. In *Proc. ICMAS-95*, pages 81–88. AAAI-Press, 1995.

[9] B. FRONHÖFER and F. KURFESS. *Cooperative Competition: A Modest Proposal Concerning the Use of Multi-Processor Systems for Automated Reasoning*. TR., Department of Computer Science, Munich University of Technology, 1987.

[10] M. FUJITA, R. HASEGAWA, M. KOSHIMURA and H. FUJITA. Model Generation Theorem Provers on a Parallel Inference Machine. In *Proc. the Int. Conf. on Fifth Generation Computer Systems*, 1992.

[11] A. GEIST, A. BEGUELIN, J. DONGARRA, W. JIANG, R. MANCHEK and V. SUNDERAM. *PVM: Parallel Virtual Machine. A Users' Guide and Tutorial for Networked Parallel Computing*. MIT Press, 1994.

[12] A. JINDAL, R. OVERBEEK and W. C. KABAT. Exploitation of Parallel Processing for Implementing High-Performance Deduction System. *Journal of Automated Reasoning*, (8), 1992.

[13] R. LETZ, J. SCHUMANN, S. BAYERL and W. BIBEL. SETHEO: A High Performance Theorem Prover. *Journal of Automated Reasoning*, (8), 1992.

[14] R. LETZ and J. SCHUMANN. PARTHEO: A High-Performance Parallel Theorem Prover. In *Proc. CADE-10*. Springer, 1990.

[15] D. W. LOVELAND. *Automated Theorem Proving: a Logical Basis*. North-Holland, 1978.

[16] E. L. LUSK, W. MCCUNE and J. K. SLANEY. *ROO - A Parallel Theorem Prover*. Technical Report ANL/MCS-TM-149, Argonne Nat. Lab., 1991.

[17] W. MCCUNE. Otter 2.0. In *Proc. CADE-10*. Springer, 1990.

[18] J. PHILIPPS. *RCTHEO II, ein paralleler Theorembeweiser*. TR., Department of Computer Science, Munich University of Technology, 1992.

[19] J. SCHUMANN. DELTA - A Bottom-Up Preprocessor for Top-Down Theorem Provers. System Abstract. In *Proc. CADE-12*. Springer, 1994.

[20] J. SCHUMANN. *SiCoTHEO - Simple Competitive parallel Theorem Provers based on SETHEO*. TR., Department of Computer Science, Munich University of Technology, 1995.

[21] J. SLANEY. *FINDER Finite Domain Enumerator Version 3.0 Notes and Guide*. TR., Australian National University, 1995.

[22] M. E. STICKEL. A Prolog Technology Theorem Prover. *New generation computing*, (2), 1984.

[23] G. SUTCLIFFE. *A Heterogeneous Parallel Deduction System*. TR., ICOT TM-1184, 1992.

[24] C. B. SUTTNER, G. SUTCLIFFE and T. YEMENIS. The TPTP Problem Library. In *Proc. CADE-12*. Springer, 1994.

[25] C. B. SUTTNER. *Competition versus Cooperation*. TR., Department of Computer Science, Munich University of Technology, 1991.

[26] C. B. SUTTNER. *Static Partitioning with Slackness*. PhD thesis, Department of Computer Science, Munich University of Technology. 1995.

4.3 Automatic Test Pattern Generation

by Andreas Ganz

Quality, reliability, and maintenance are key points in the production of digital integrated circuits (ICs) and boards. In combination with the continuing move towards higher levels of integration, this leads to an increasing importance of the test phase. The task of Automatic Test Pattern Generation (ATPG) is to obtain a set of logical assignments to the inputs of a circuit that will distinguish between a faulty and a fault-free circuit during production line test. The gate-level circuit model, where the basic primitive is a logic gate, is commonly utilized for ATPG, since this level of abstraction combines sufficient accuracy with acceptable computational expense. Various fault models map physical defects to fault effects on the gate-level. They can be split into two groups, so-called static (DC) fault models and dynamic (AC) fault models. The most popular static fault model is the stuck-at fault model, where a signal is supposed to be constant at a logic value. For modelling dynamic defects, the gate- and path-delay fault model are wide-spread. Within this contribution we focus on the path-delay fault model, where delay faults are modelled along a sequence of successive signals from an input of a circuit to an output.

Independent of the applied fault model, the two following requirements occur during test pattern generation for each fault:

Fault excitation: On the faulty signal, the inverse value to the assumed fault has to be adjusted.

Fault propagation: The circuit must be sensitized in a manner such that differences between the fault-free and the faulty circuit can be observed at an output.

Internal signal assignments resulting from these two tasks have to be justified by input signals. This problem can also be viewed as a *boolean satisfiability decision problem* [13], inherent in numerous other tasks. *Branch-and-Bound* search techniques, together with a variety of heuristics for pruning the search trees, are the basis for many popular techniques [8; 5; 16], solving the satisfiability problem in ATPG.

ATPG is known to be NP-complete even for combinational circuits and hence it is very difficult to speed up the test generation process by backtracking mechanism during search space traversal. Nevertheless, efficient heuristics to speed up test generation have been proposed. But handling the increased complexity, caused for example by VLSI circuit design and enhanced fault models, has been severely limited by the restrictions of conventional ATPG tools on general purpose computers. Therefore, as in many other areas of CAD, parallelism is utilized to decrease both computation time and memory requirements. An excellent and detailed summary of parallel methods and algorithms in electronic design automation can be found in [2].

4.3.1 Parallel Test Pattern Generation

There are several types of parallelism inherent in ATPG: fault parallelism, search parallelism, heuristic parallelism, and circuit parallelism. Fault parallelism refers to dealing with different faults in parallel. Search parallelism investigates disjoint subtrees of a decision tree concurrently in a parallel branch-and-bound search. In heuristic parallelism one fault is treated, using different heuristics in parallel. A partitioning of the circuit is the basis for circuit parallelism. Furthermore, there exist methods, which don't fit into one of these groups. The two techniques, which turned out to be most efficient, are fault parallelism and search space parallelism. Thus, we focused on both of these two methods.

Our described test generation environment consists of two tools. The first one handles stuck-at faults in sequential circuits and is called PESSENTIAL. It is based on the state-of-the-art test pattern generator ESSENTIAL [17]. Within this tool, fault parallelism is applied as introduced in the next section. For certain stuck-at faults in sequential circuits, called *hard-to-detect faults*, ATPG has to proceed a huge search space and therefore takes a long time. For these faults, PESSENTIAL splits up the search space and works on the resulting disjoint regions in parallel. The second tool performs ATPG for path-delay faults in combinational blocks and is called PTIP, based on the test design tool TIP [10]. Since PTIP applies fault parallelism, we introduce it in conjunction with the fault parallel method of PESSENTIAL.

A NOW (Network Of Workstations) connected via Ethernet is used as platform for running our distributed ATPG tools. In order to avoid too much communication on the net, a central process, called client, is applied for coordination of the entire system. The other processes, which are receiving tasks from the client, are called servers. This organization of processes leads to a star topology, as mentioned in table 4.5 on page 157.

The client is responsible for partitioning the whole ATPG-task into single subtasks, i.e. building fault set partitions or splitting the search space. These subtasks are distributed to processes, which have finished their previous subtask and sent back the respective results. When assigning the computation subproblems, the location of the process is not considered. These facts are summarized at the bottom of table 4.5 on page 157. From this organization it can be seen, that the decision structure is centralized. Before describing our load balancing mechanisms in more detail, our tools are introduced.

4.3.2 Fault Parallelism

Fault parallelism is a simple but efficient approach, based on partitioning the fault list. Each processor performs ATPG as usual for a subset of the fault dictionary. The way how faults are distributed influences the overall performance in a very sensitive way. In order to achieve a high speedup it is obligatory to maintain a good load balance and to consider constraints due to dependencies between single faults.

Dependent faults are detected by the same test sequence. Fault simulation between ATPG runs is used for recognition of these faults. For additionally detected faults, no ATPG has to be performed. This considerably speeds up the entire process, since ATPG is more complex than fault simulation. Depending on the manner of distributing the fault set partitions, dependent faults can be located on different processors. Therefore, test sequences for dependent faults can needlessly be generated concurrently. Various static partitioning methods tackle this problem. A more detailed discussion of dependencies between faults and the applied partitioning methods can be found in [12].

Dependencies between faults mainly influence the overall running time of ATPG, whereas the different computational complexity for hard-to-detect faults and easy-to-detect faults results in a significant load imbalance. Different computation times for these faults can only be approximated by static partitioning approaches. This has to be overcome by assignment of fault partitions as well as adaptive repartitioning of fault partitions.

In the following, fault parallelism for both stuck-at faults in sequential circuits (PESSENTIAL) and path-delay faults in combinational blocks (PTIP) are introduced. Specific partitioning and assignment approaches, which consider load distribution aspects, as well as supplementary dynamic repartitioning is discussed.

4.3.2.1   Fault Parallelism for Stuck-at Faults in Sequential Circuits

a) Fault Set Partitioning

As already mentioned, fault set partitioning is based on building disjoint subsets out of the entire fault set, which are well suited for being treated concurrently on different processors, i.e. a good load balance can be obtained and the computational overhead is low. For our fault set partitioning method, which is explained more detailed in [12], dependencies between faults are examined. Mainly, the following two different partitioning methods were investigated.

Test Counting (TC)

The *Test Counting* algorithm for combinational circuits was introduced by Akers and Krishnamurthy [1]. Test Counting means that the number of test vectors, needed for testing a set of faults, is counted. This approach also results in a set of faults, which are pairwise independent. These faults can be selected as candidates for fault set partitioning. Characteristic for Test Counting is that no fault simulation is used. Instead, constraints are made for each signal as well as for the gates of the circuit, which are formulated as conditions, leading to a large system of inequalities. By solving this system, a lower bound of necessary test vectors and a set of pairwise independent faults is determined.

Since the resulting set of independent faults is relatively small, it is not possible to use the capacity of large parallel systems sufficiently. In order to get more independent faults, and hence a better load distribution, it is useful to run Test Counting several times, removing previously determined independent faults. Furthermore, Test Counting is not limited to a preprocessing phase but can be performed multiple times during ATPG. Since more independent faults are needed for a larger number of processors in order to achieve a sufficient load balance, the repetition rate increases with the size of the parallel system. Therefore, the quite high computational overhead caused by this approach limits the speedup. The sequential Test Counting algorithm is a suited technique for fault set partitioning in sequential circuits, but for practical use with a large number of processors the performance limitations are significantly.

Dominant Fault Collapsing (DFC)

Dominant Fault Collapsing is based on two relationship between faults, called equivalence and dominance. Two faults are called equivalent, if test sequences always detect both faults. Since those faults can never be distinguished, only one of them has to be tested. A fault ν dominates another fault μ, if a test sequence testing μ always tests ν, too. Therefore, only for the dominated fault a test sequence has to be generated.

An adequate structural method to extract dominance and equivalence relationships in sequential circuits is Dominant Fault Collapsing [11]. A directed graph represents all modelled stuck-at faults and the relations between faults. With the help of this graph a set of faults is created, such that all dominant and equivalent faults are tested implicitly.

Figure 4.10 shows the speedup for different partitioning methods with respect to the number of processors. The curve TC shows the progress when applying test counting algorithm. For only few processors, an improvement is achieved compared to the curve without an enhanced partitioning

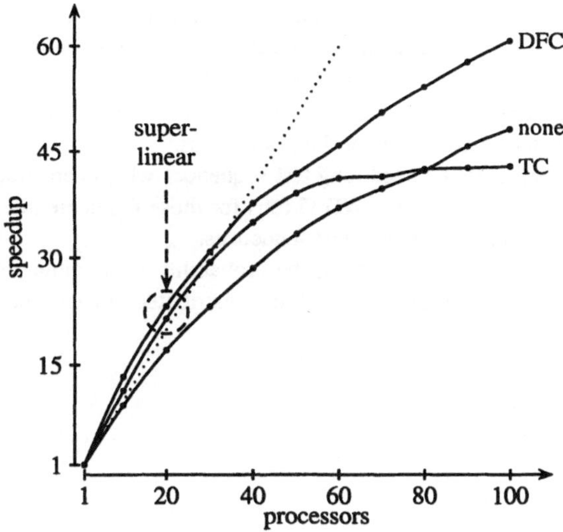

Figure 4.10: Speedup for s3271 with different partitioning methods

approach. For a larger amount of servers, the calculation overhead even leads to a worsening of performance. In contrast, the proposed fault set partitioning by dominant fault collapsing is clearly superior since selection of suited faults leads to a good load balance.

b) Fault Sorting
In order to achieve a good load balance by fine granularity, only one target fault is sent to a server process each time. This also decreases the probability for generating test sequences for compatible faults concurrently. Two faults are called compatible, if they are neither equivalent nor dominant, but they are detected by at least one common test sequence. Sending single faults is applicable since due to the fault sorting algorithm the computation time for a fault is usually long compared to the communication overhead.

As mentioned earlier, the required ATPG time for a target fault cannot be determined exactly during a preprocessing phase. A heuristic named COP [3] is used, which evaluates the circuit structure, in order to estimate the expense for generating a test sequence for a fault. COP determines two probabilities for each signal of a circuit. The first one is called controllability C_s and gives the probability that a certain value at signal s can be excited. Second, the probability that a signal s can be observed at one of the circuit outputs is given by its observability O_s.

To detect a stuck-at fault at a signal s, the value inverse to the fault has to be excited and the signal has to be observable. Under the assumption, that controllability and observability are independent of each other, the probability for testing a fault on the signal s by a random test sequence can be given by:

$$C_s \cdot O_s \quad \text{for } s/0, \text{ and}$$
$$(1 - C_s) \cdot O_s \quad \text{for } s/1 .$$

(4.1)

If this probability is low, it is assumed that treating this fault is difficult and therefore the computation time for ATPG is high. Thus, faults are sorted in a manner, that faults with a low detection probability are handled first. Advantages of this method are:

- Long computations for hard-to-detect faults don't occur at the end, where they may significantly influence the termination time of the last process.
- Generally, hard-to-detect faults need long test sequences which can additionally test many easy-to-detect faults, too. Therefore, ATPG runs for those faults are not necessary, leading to both less computation time and less test sequences.
- With increasing length of test sequences, the probability of a compatible fault decreases. Hence, also the probability of multiple work in a distributed environment decreases.

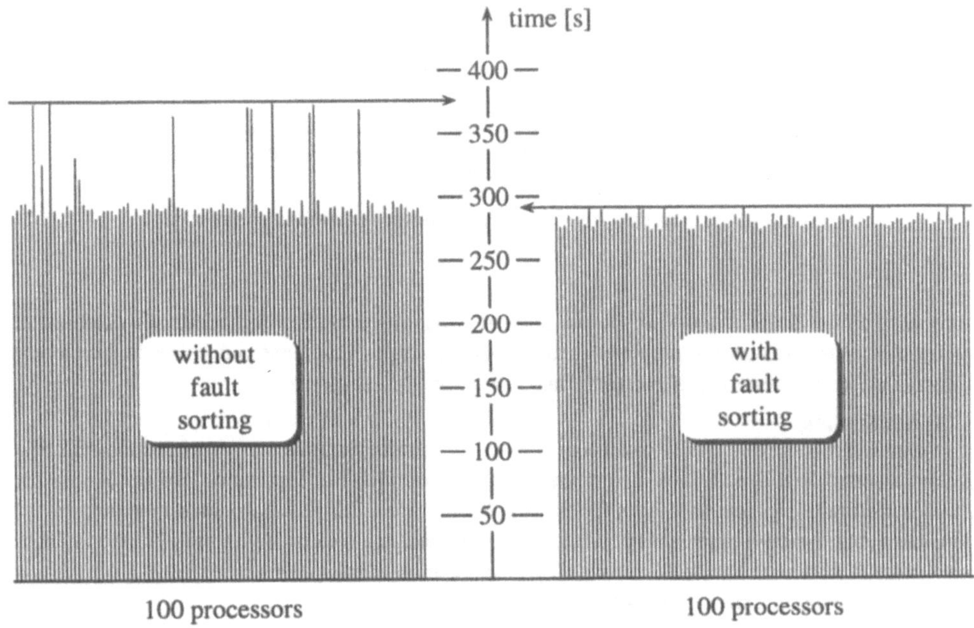

Figure 4.11: Load balance for s9234.1 by fault sorting

Figure 4.11 shows the achieved load balance by fault sorting for circuit s9234.1, where each vertical line represents the activity of a server. If no fault sorting is applied, the situation illustrated in the left part of figure 4.11 occurs, in which some stragglers prevent termination of the entire ATPG process. In the right part of the figure, all 100 servers finish at approximately the same time, because the longest tasks start at the beginning. Therefore, sorting decreases the overall computation time from 376 to 290 seconds and increases the speedup from 71 to 92.

As described above, the functionality of load balancing for fault parallelism for stuck-at faults consists of both partitioning and assignment, whereas no explicit migration takes place. In order to avoid load imbalance, small partitions are built which, in conjunction with fault sorting and respective assignment, lead to a good load distribution. A more detailed explanation is given in section 4.3.4 on page 157.

4.3.2.2 Fault Parallelism for Path-Delay Faults in Combinational Blocks

The increasing complexity of digital logic systems and the continuing move towards higher levels of integration has steadily augmented the importance of dynamic testing. A dynamic fault model, which considers delay faults distributed in the circuit, is the path delay fault model. One major disadvantage of the path delay fault model is the huge size of its fault dictionary, since the number of paths in the circuit may grow exponentially with the circuit depth. More detailed information about the path delay fault model can be found in [6; 7].

a) Fault Set Partitioning

Some experiments were carried out in order to investigate if it is possible to transfer some partitioning strategies from the stuck-at fault model to the path delay fault model. The following paragraphs show, why different strategies have to be applied for the path delay fault model.

It has been examined how many faults are detected by one test pattern on the average, which gives us information about the extent of dependencies between faults. The degree of dependencies is by far less for the path delay faults than for stuck-at faults, i.e. for the majority of the faults, test generation has to be performed. As a consequence, forming larger partitions does not cause much redundant computations.

As in the stuck-at fault model, computation times for faults differ in a very wide range. The reason for this is a two-phase approach for generating test patterns. In the first phase, all side inputs of gates along a targeted path are set to values, such that a transition on it can be propagated. From this signal values, all possible implications are carried out. If a conflict arises in this phase, no patterns exist to test the path. This kind of faults is processed very fast. If no conflict arises, in a second phase all internal signal values have to be justified by explicit value assignments on circuit input signals. Thereby, a considerable part of the huge search space has to be traversed for hard-to-detect faults. Therefore, test generation for these kind of faults may take a long time. There exists no suited heuristic to estimate the computation time for generating a test pattern for a target fault.

It is not applicable to send single faults for ATPG because of two reasons. First, faults are worked on concurrently by using bit-parallelism, i.e. typically 32 or 64 paths can be tackled at one time by exploiting the machine word length. This advantage would be lost when working on only one target fault. Second, for minimization of the expense of sensitization, the fact is used that two neighboring path have almost all path segments (parts of paths) in common. These segments have only to be sensitized once, if neighboring paths are worked off subsequently. Thus it is advantageous to work off neighboring paths subsequently on one processor.

Furthermore, the occurrence of faults with a certain computational expense is clustered, i.e. often several hundred paths are detected as being untestable at the same time, because of having the same untestable path segments. Since these faults are only detected during test generation, it is not possible to cluster them within one partition in a preprocessing partitioning step.

All the above discussed facts form the motivation to create relatively large fault set partitions, typically containing several hundreds of paths. This is applicable since large circuits, for which parallelism is senseful, consist of several millions of paths. In order to tackle the already mentioned problem of different computational complexity of faults and their clustered occurrence, dynamic load balancing mechanisms based on repartitioning as described in the following are applied.

b) Repartitioning

Repartitioning is carried out, when at least one server has finished ATPG for its assigned fault partition and no further initially created, unprocessed partition is left that could be assigned. Then, the server which is expected to have the highest load, is asked to split up the untackled part of its fault partition. Up to now, the best heuristic was to count the number of unprocessed faults on each server, which implicitly considers the progress of ATPG. This results in a single dimensional and application defined load index. The aggregation time of the index is instantaneous, since after every few ATPG runs the current number of remaining faults is sent to the client.

The determined server forms two new partitions. These are created by equally sized bipartioning of the unprocessed part of its currently tackled fault partition. The first of the newly created partitions is processed on the server, which has done the partitioning, and the second on one of the idle servers. The repartitioning is repeated as long as one server is idle or the estimated remaining time of the slowest server is under a given limit. The migrant for this load balancing step is computation (ATPG) for a newly created fault set partition. The migration set size is limited, since fault sets are finite, of course. Our ATPG tools are currently running on a homogeneous NOW. Thus, our migration mechanism has not to consider heterogeneity, but the necessary extensions are marginal. No data have to be transfered before the actual migration. Therefore, the pre transfer media is not applied, as shown in table 4.6 on page 158. Migration initiation of this approach is delayed, because the sending server has to finish its current ATPG run for target faults before splitting its fault partition. The respective transfer policy is complete, since fault partitions contain all data necessary for ATPG.

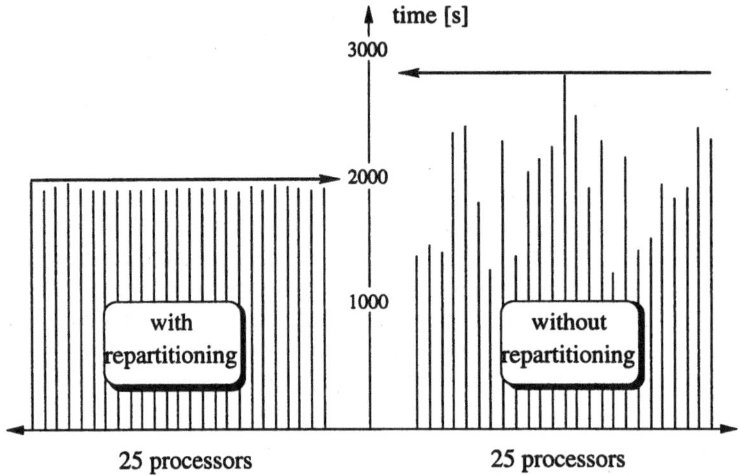

Figure 4.12: Load balancing by repartitioning

Figure 4.12 shows the typical improvement for a benchmark circuit, which has a large load imbalance without repartitioning. Load imbalances could also be tackled by building smaller initial partitions, but this leads to considerable computational overhead inherent in ATPG, as already mentioned above. Since the computational overhead of dynamic repartitioning is very small, larger initial partitions are built and repartitioning is applied if imbalance is detected.

4.3.3 Search Space Parallelism

Usually, test generation is a depth-first search, guided by heuristics. For hard-to-detect faults, a huge part of the search space has to be processed, which increases the computing time requirements considerably. This can cause an abortion of the search, where the number of canceled decisions, called backtracks, is usually taken as the criterion for abortion. A promising method for obtaining tests for hard-to-detect faults is to start with a breadth-first search and switch to a depth-first search later on. As the amount of data, which has to be processed, is too high for a single processor, parallel systems can be employed successfully for these faults.

a) Size of the Search Space

After fault effect propagation, FAN-based [4] ATPG tools use a backtrace procedure to drive requirements for unjustified signal values through the combinational part of the circuit from gate outputs to gate inputs. These requirements have to be satisfied by explicit value assignments, which are made at a primary input, a pseudo primary input (output of a flip-flop), or a certain fanout stem (head line).

In our case, the logic values 0, 1, or U are assigned. The alternative U describes the condition that a signal depends on the output of a flip-flop with an undefined state. Successive decisions are maintained in a *search tree*, also called *decision tree*, where each further decision adds one level more. Figure 4.13 displays a graphical representation.

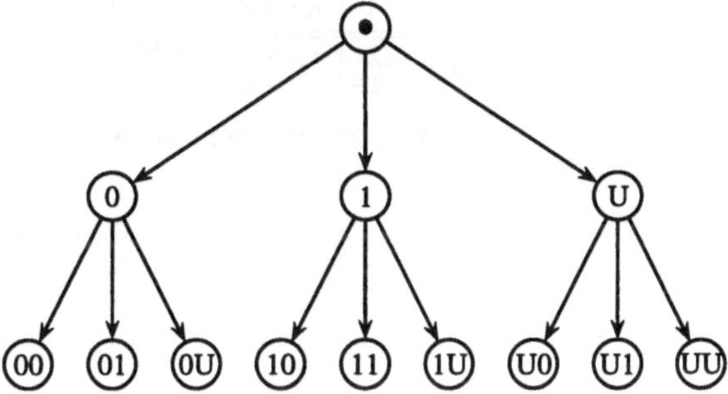

Figure 4.13: Search tree

A complete traversal of the shown decision tree from left to right by a depth-first backtrack algorithm is given in figure 4.14 . Of course, the order of assignments could also be $1 \rightarrow 0 \rightarrow U$ at any time. Usually, testability measures are applied to decide whether to try 0 or 1 first. The alternative tried next is always the inverse value of the first, and U is eventually tried last. Let n be the maximal number of decisions in a path from the root node to a leaf node in the decision tree. In a worst-case estimation, the breadth of the decision tree grows with 3^n. Table 4.4 gives the maximal possible size of the search space for some circuits to show its huge dimensions.

Figure 4.15 shows the search space traversal for a certain fault in circuit s6669. Compared with table 4.4 , the average depth of 16 is significantly smaller. This happens because a large part of the search space can be pruned due to contradictory signal assignments. Nevertheless, it is not unusual that even allowing 10 000 backtracks does not lead to a solution. If the inverse value at the fourth decision step in this example would have been used, however, the solution were found without any backtrack.

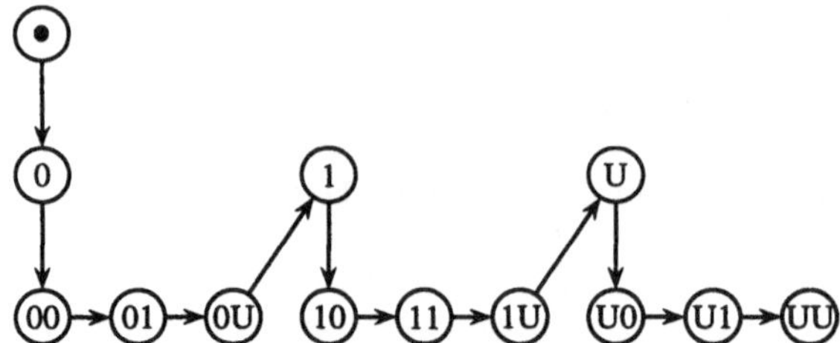

Figure 4.14: Depth-first search

circuit	max. depth	max. breadth
s526n	1878	$1.0807 \cdot 10^{896}$
s838.1	98	$5.7264 \cdot 10^{46}$
s1488	285	$9.5402 \cdot 10^{135}$
s3271	84	$1.1972 \cdot 10^{40}$
s3330	248	$2.1187 \cdot 10^{118}$
s3384	134	$8.5950 \cdot 10^{63}$
s6669	782	$1.2847 \cdot 10^{373}$

Table 4.4: Maximal size of search space

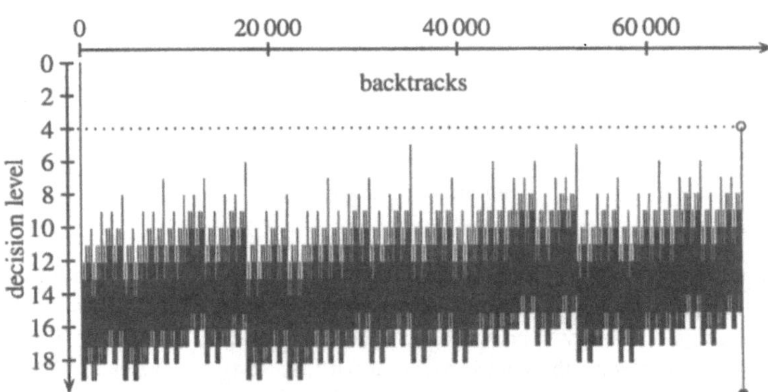

Figure 4.15: 70 000 backtracks

Performing a rough estimation, after $3^{16-4} = 531441$ backtracks the wrong decision will be taken back and the inverse value leads to a solution. In fact, after 68949 backtracks the inverse value is assigned and a test sequence is found immediately, marked by the symbol "o", the symbol "•", and the dashed horizontal line in figure 4.15. Nevertheless, such a high backtrack limit is not reasonable. Therefore, sequential ATPG for hard-to-detect faults is generally not solved satisfactory.

b) Dynamic Partitioning of the Search Space

In contrary to other search space partitioning methods, e.g. [14; 15], a new strategy combining breadth- and depth-first search is proposed, which is especially useful for hard-to-detect faults as already mentioned. Our approach rates **all** possible assignments at each decision node up to a chosen depth bound. As a consequence, that node out of the whole area up to the specified depth bound is selected, which has the best rating. For rating the nodes testability measures are used, which generally give good results in ATPG. For our application, SCOAP [9] performed generally better than COP, which has already been described.

SCOAP estimates the controllability C_s as the minimum number of necessary assignments in the circuit to justify the given logic value of a signal s. The observability, denoted by O_s, is related to both the distance of s to a circuit output and the controllability of assignments required to propagate the logic value of s to a circuit output. A lower value means a better controllability or observability, respectively. Assuming independency between controllability and observability, a measure $M(i)$ for each single decision step i is obtained by evaluation of:

$$C_s(0) \cdot O_s \quad \text{for assignment of 0,}$$

$$C_s(1) \cdot O_s \quad \text{for assignment of 1, and} \tag{4.2}$$

$$\frac{1}{2} \cdot \left[C_s(0) + C_s(1) \right] \cdot O_s \quad \text{for assignment of U.}$$

The measures of successive decisions are summed up. Hence, a deeper nested decision has probably a worse rating. Thereby, a preference for a breadth-first exploration of the maintained search tree is gained. Let n be the number of a decision step, then the rating $R(n)$ of this step is obtained by

$$R(n) = \sum_{i=1}^{n} M(i) \ . \tag{4.3}$$

Due to the exponential growth of the decision tree a depth bound is used also. Below the depth bound no rating takes place and a normal sequential ATPG is performed.

For each fault, the client builds up the search tree containing the rating for each decision. The servers perform the actual ATPG. As soon as a server has finished its startup procedure, it queues up for requesting work from the client. The central administration avoids problems appearing in [14], where the idle processor has to decide from which active server it requires work, leading to additional communication and a worsening of load balance.

The first server in the queue gets a partition of the search space for the target fault, starts ATPG and reports back to the client up to a given depth bound every decision including changes due to backtracks. Thereby, the client is able to maintain a decision tree and monitor which region of the search space the servers are currently processing. For each request, the client scans the unprocessed decisions for the best rating and sends the server a decision vector leading to the start node, from where the server begins to perform the task described above.

With help of figure 4.16 , the partitioning of the search space according to our algorithm will be explained. The numbers beside each node give the ratings $R(n)$ for the respective decision step n. For simplicity, we choose successive numbers and refer to a node with a rating of n as node "n". To the first three idle processors in the queue, the nodes ranked with 1, 2, and 3 will be assigned. This leads to a breadth-first search up to a given depth bound as already mentioned. The next requesting idle processor will get the decision node "5", as the processor starting at node "1" already works on the search space beyond node "4".

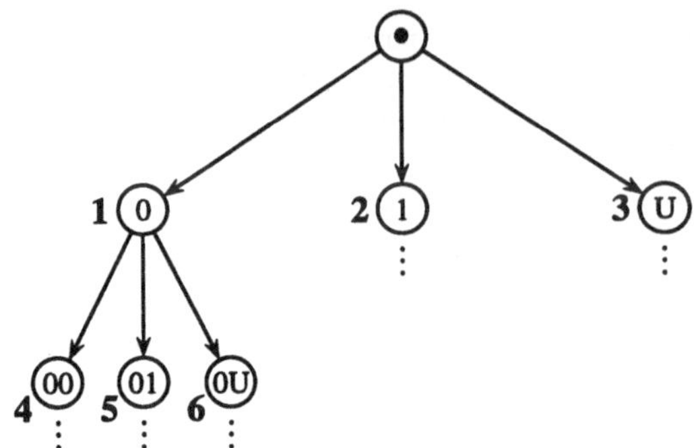

Figure 4.16: Combination of depth- and breadth-first search

Obviously, the processor started on decision node "1" has to be stopped if it backtracks from decision node "4" to decision node "5". Therefore, two regulations are applied to prevent a server from processing a region which has been assigned to another server. First, a server never backtracks beyond its start node. Second, every backtrack performed above the depth bound is confirmed by the client, which checks that no other server is processing this region. In both cases the server terminates its current task and the client marks the region as containing no solution.

A normal backtrack restriction is applied to limit the search. Within this limit, the server may find a test sequence, or every decision leads to an inconsistency, if no solution exists in that region. When hitting the backtrack limit, the server aborts the search. In each of these cases, the server sends a specific message to the client. Afterwards it terminates the current task and queues up in the row of idle servers.

If every region of the search space has been processed and none of them contained a solution, the fault is proven to be not testable. However, one abort is enough to prevent this proof. On the other hand, if a server finds a test sequence, the client aborts all other servers processing this fault. As the successful server has the test sequence it also performs fault simulation, whereas the other servers continue with processing of the next fault. Therefore, no time expensive synchronization after each fault is necessary. Based upon the procedure described so far, one hard-to-detect fault after the other is processed.

The concept of requesting servers helps to achieve not only a dynamic partitioning but also a dynamic load balancing, suited well for the sudden load changes in a network environment. By means of adjustment of the depth bound, a compromise between granularity and communication overhead can be achieved. For a deep bound, there exist many smaller partitions and thus a good load balance due to fine granularity is probable. On the other hand, processors send more messages and a lot of organization has to be done by the client. Therefore, in order to prevent the client from becoming a bottleneck, a rather small depth bound is chosen.

4.3.4 Summarizing the load balancing mechanisms

Concluding, the classification scheme, which is proposed in this publication, is applied to our ATPG tools as explained above. Most of the classifications are valid for all of the above described ATPG tools. If no remark is made according to a special fault model or method of parallelism, the tackled characteristic is valid for all different approaches.

Table 4.5 concludes the facts, described in the following. The client process performs our domain specific load balancing algorithms, which can be divided into static load balancing methods, e.g. suited partitioning of the fault list, and dynamic load balancing methods, e.g. repartitioning of the fault list during ATPG. Dynamic load balancing algorithms are adaptive, i.e. they implicitly consider other influences, e.g. load imbalance caused by other processes and heterogeneous hardware. Furthermore, our load distribution method provides the functionality of both partitioning and assignment. The integration level of load balancing is completely on the application level for all the aspects according to table 4.5, i.e. monitoring, strategy, and mechanisms. As a consequence, migration transparency and source code transparency is not applicable. The general structure of the load balancing scheme inherent in our ATPG tools were already described in section 4.3.1 on page 146.

I. General Classification

Objectives	
domain specific	yes
intent	application adaptive
function	partitioning & assignment
Integration Level	
monitoring	application
strategy	application
mechanisms	application
compiler support	none
Structure	
problem structure	location independent
subproblems	computation
workspaces	none
targets	processes

II. Strategy Classification

System Model	
Model Flavour	client/server
Processor Topology	NOW
Appl. Topology	star
Transfer Model	
Transfer Space	systemwide
Transfer Policy	non-preemptive
Information Exchange	
Information Space	central
Information Scope	complete
Coordination	
Decision Structure	centralized
Decision Mode	cooperative
Participation	partial
Algorithm	
Decision Process	static & dynamic
Initiation	central
Adaptivity	fixed
Cost Sensitivity	none
Stability Control	not required

Table 4.5: General and Strategy Classification of ATPG

Next, strategy classification as shown in table 4.5, is summarized. The considerations leading to a star topology were already described in section 4.3.1 on page 146. The applied star topology has impact on numerous further strategy classification aspects. So, the client centrally holds the complete information. Hence, decisions are made solely by the client, i.e. the decision structure and initiation is centralized and the decision mode is cooperative.

The decision process can either be static or dynamic. In case of fault parallelism, fault set partitions are created statically. These initial partitions are sent to each idle server. Hence, transfer space is systemwide and transfer policy is non-preemptive. This holds also for repartitioning of the fault list as explained in section 4.3.2.2 on page 151 as well as dynamic partitioning of the search space, which forms the dynamic decision process. For both static and dynamic decision process, participation is partial and adaptivity is fixed. The costs caused by load balancing during ATPG is not considered, since they are negligible. Because all tasks can be worked off independently, stability control is not necessary.

Now, classification of load models of different methods of parallelism and fault models is discussed. The load index is generally application defined and single-dimensional. It is useful to distinguish between load determination for receiving and sending entities. The load of the receiving server is always a binary value, i.e. the server is idle or busy. In case of fault parallelism for stuck-at faults, the sending entity is the client and hence no index determination is performed, since it sends the next task for each idle server. For fault parallelism for path delay faults as well as search space parallelism, a dynamic partitioning takes place, and thus the sending server has to be determined. This is done by estimating the remaining work for path delay faults, as explained in section 4.3.2.2 on page 151, and by taking the unprocessed decision with the best rating for search space parallelism, which are both implicitly considering the load by progress of the task. The index propagation is after measurement and the model adaptivity is fixed. Independent of the method of parallelism, aggregation time is instantaneous and aggregation space is application related. Since index dimension is single-dimensional, index combination and weight selection is not applicable.

<table>
<tr><td colspan="2" align="center">III. Load Model</td></tr>
</table>

Characteristic	Categories
Load Index Properties	
index dimension:	single-dim.
index type:	appl. defined
aggregation time:	instantaneous
aggregation space:	appl. related
Load Value Composition	
index combination:	none
weight selection:	not applicable
Model Usage Policies	
index measurement:	implicit
index propagation:	adaptive
model adaptivity:	fixed

IV. Migration Mechanisms

Migration Mechanisms	
Migrant	computation
Migration Set Size	limited
Heterogeneity	no
Migr. Initiation	immediate/delayed
Pre Transfer Media	none
Compression	no
Transfer Policy	complete
Residual Deps.	none
Migration Transp.	no
Source Code Transp.	no

Table 4.6: Load Model and Migration Mechanisms of ATPG

For fault parallelism for stuck-at faults, no actual migration takes place, since initially created partitions are distributed among idle servers. In search space parallelism, the idle server starts search space traversal at the unprocessed node with the best rating immediately. The other servers are prevented from processing this region of the search space by the client as described in section 4.3.3 on page 153. For the path delay fault model, computation entities, which are limited in their migration size, are migrated between servers after a short delay , as described in section 4.3.2.2 on page 151.

The following attributes are valid for all of our parallel approaches. As already mentioned, no heterogeneity is currently implemented, but only a few extensions would be necessary since the tools are already ported to different hardware platforms. Because the amount of data and the respective required time is small compared to computation time, no data compression is applied, transfer policy is complete, and no pre transfer media is used. As all subtasks are independent ATPG runs, residual dependencies do not exist.

References

[1] S. B. AKERS and B. KRISHNAMURTHY. Test Counting: A Tool for VLSI Testing. *IEEE Design & Test of Computers*, pages 58–73, October 1989.

[2] P. BANERJEE. *Parallel Algorithms For VLSI Computer-Aided Design*. Prentice Hall, 1994.

[3] F. BRGLEZ. On Testability Analysis of Combinational Networks. In *IEEE Int. Symp. on Circuits and Systems (ISCAS)*, pages 221–225, June 1984.

[4] H. FUJIWARA and T. SHIMONO. On the Acceleration of Test Generation Algorithms. *IEEE Trans. on Computers*, 32(12):1137–1144, December 1983.

[5] H. FUJIWARA. *Logic Testing and Design for Testability*. Computer Systems Series. The MIT Press, Cambridge, 1985.

[6] A. GANZ, P. TAFERTSHOFER and H. WITTMANN. Parallel Computation of Delay Fault Probabilities. Technical Report TUM-LRE-96-2, Technical University of Munich, May 1996.

[7] A. GANZ, H. WITTMANN and M. HENFTLING. Parallele Berechnung von Verzögerungsdefektwahrscheinlichkeiten. In *ITG/GI-Workshop Testmethoden und Zuverlässigkeit von Schaltungen und Systemen*, March 1996.

[8] P. GOEL. An Implicit Enumeration Algorithm to Generate Tests for Combinational Logic Circuits. *IEEE Trans. on Computers*, 30(3):215–222, March 1981.

[9] L. H. GOLDSTEIN. Controllability/Observability Analysis of Digital Circuits. *IEEE Trans. on Circuits and Systems*, pages 685–693, September 1979.

[10] M. HENFTLING, H. WITTMANN and K. J. ANTREICH. A Formal Non–Heuristic ATPG Approach. In *European Design Automation Conf. with EURO-VHDL (EURO-DAC)*, pages 248–253, September 1995.

[11] P. A. KRAUSS and M. HENFTLING. Efficient Fault Ordering for Automatic Test Pattern Generation for Sequential Circuits. In *Asian Test Symp.*, pages 113–118, November 1994.

[12] P. A. KRAUSS. *Parallelisierung der automatischen Testmustergenerierung in sequentiellen Schaltungen*. Ph.d. thesis, Technical University of Munich, Jan 1995.

[13] G. D. MICHELI. *Synthesis and Optimization of Digital Circuits*. McGraw–Hill, Inc., New York, 1994.

[14] S. PATIL, P. BANERJEE and J. H. PATEL. Parallel Test Generation for Sequential Circuits on General–Purpose Multiprocessors. In *ACM/IEEE Design Automation Conf. (DAC)*, pages 155–159, June 1991.

[15] B. RAMKUMAR and P. BANERJEE. ProperCAD: A Portable Object-Oriented Parallel Environment in VLSI CAD. In *IEEE Int. Conf. on Computer Design (ICCD)*, pages 544–549, 1992.

[16] M. H. SCHULZ, E. TRISCHLER and T. M. SARFERT. SOCRATES: A Highly Efficient Automatic Test Pattern Generation System. In *IEEE Int. Test Conf. (ITC)*, pages 1016–1026, September 1987.

[17] M. H. SCHULZ and E. AUTH. ESSENTIAL: An Efficient Self-Learning Test Pattern Generation Algorithm for Sequential Circuits. In *IEEE Int. Test Conf. (ITC)*, pages 28–37, August 1989.

4.4 MPSIM - **Parallel Event Driven Simulation of Logic Circuits by** *Time Warp*

by Rolf Schlagenhaft

MPSIM (Multi Partition Simulator) [13; 14] is a distributed discrete event simulator for logic circuits at gate level, running on a network of workstations (NOW). It is synchronized by the optimistic *Time Warp* protocol [10], which takes best advantage of parallelism of the simulated model compared to other mechanisms. But *Time Warp* is more sensitive to load imbalances than other parallel applications, because its total CPU time usage is not constant like in many distributed algorithms, but can highly increase in badly balanced situations. Most simulators based on *Time Warp* are only balanced statically at startup time and hence yield good results on hardware, which may be used by the simulation system exclusively. Unfortunately, this is not the usual situation in workstation clusters at circuit design labs. To make a *Time Warp* simulator a suitable tool for designers, it definitely needs a dynamic load balancing mechanism. MPSIM solves this problem **domain specific** and is based on the former simulator PSIM (Parallel Simulator) [1; 2; 3; 4].

4.4.1 Parallel Discrete Event Simulation (PDES)

Discrete Event Simulation

In a system for discrete event simulation (DES) the static information about the physical system (topology, behavior of components) to be simulated is coded in the simulation model. During a simulation run, the simulator advances in simulation time by evaluating the model components and thereby generating events. The event administration holds them until their execution time is reached. The execution of events again leads to state changes by model evaluations and mostly to the generation of further events. The steps event execution, model evaluation and event generation are called simulation cycle. Simulation time is represented implicitly by the execution time of the actual event. Since events are only generated for future simulation times (causality of simulation model), simulation time never decreases in sequential discrete event simulation.

Time Warp

Time Warp is a protocol for synchronizing a parallel distributed discrete event simulation (PDES) system consisting of several simulators and a controlling instance. Compared to normal DES, additional problems and aspects arise. Here, *Time Warp* is only outlined as far as necessary for understanding of the following sections. More detailed descriptions can be found in [10; 2; 3; 4].

In *Time Warp* , distribution is done by partitioning and distributing the model to several machines (see section 4.4.2.1). The participating simulators interact by exchanging events, which are relevant to other simulators. The simulation time clocks are not synchronized directly, but implicitly by the execution times of the transmitted events. Therefore, there is no common simulation time, which is valid for all simulators as in sequential DES, but every simulator has its own clock. This optimistic synchronization approach uses the **virtual time** paradigm [10]. Each simulator may advance in its own **local virtual time (LVT)** as fast as possible and without taking the speed of the other simulators into consideration. This may lead to a situation, where a simulator A receives an event from simulator B, which has an execution time less than the local virtual time of A. Such events are called **straggler events**. The advance in virtual time of simulator A was too optimistic and the results concerning simulation times greater than execution time of the straggler event are potentially wrong. In *Time Warp* they can be undone by the state restoration technique called **rollback**. Every simulator stores and maintains old states or state changes. The two basic methods are checkpointing [6; 8] and incremental state saving [12]. In Mpsim , an the effective incremental method described in [3] is applied. Of course, it is not possible to hold all states, which occurred during a simulation run until its end, because memory is limited. The following mechanism detects and deletes obsolete state information, that will not be used any more and thereby limits memory consumption for saved states to a certain size, independent of the length of the simulation run. Because of causality of the modeled system, there is no need to rollback to a virtual time less than execution times of all untreated events in the whole simulation system. This minimum of execution times in the *Time Warp* system is called **global virtual time (GVT)**. Therefore, only state information, which is needed to restore states of virtual time GVT or greater must be kept. All other state information can be deleted. Unfortunately, GVT cannot be calculated exactly in a PDES system on distributed hardware. Several approximation techniques were suggested [2; 7; 11] to determine a lower bound. In Mpsim GVT is calculated asynchronously [4].

4.4.2 Load Balancing within Mpsim

The turnaround time between subsequent design steps is very important in circuit industry. In design labs a reasonable part of time is used for circuit simulation. In order to reduce turnaround time, the minimization of real time until termination of a simulation run has high priority. Therefore, the goal of the load balancing approach in Mpsim is not minimization of any resource utilization but maximum simulation speed.

4.4.2.1 Entities

In general, we have two possibilities of looking at distributed discrete event simulation and finding suitable entities for load balancing.

Events: One possibility is to use events as entities. Events have a relatively small lifetime from there generation to their execution. The execution itself lasts very short compared to the total simulation time. These features could lead to the assumption, that discrete event simulation belongs to the location independent class of problems. Each event could be interpreted as an entity and assigned to a certain target for administration and execution.

But this approach does not take into account, that for the execution of an event, model data is needed, too. If load balancing would be implemented based on that location independent model of DES, we could observe a huge amount of data traffic, since events for the same part of the model are simulated in different targets.

Model data: The second way of looking at DES assigns it to the **location dependent** class of problems by interpreting parts of **model data as entities**, not events. This view of the problem corresponds to the *Time Warp* algorithm already used in Psim , which is outlined in section 4.4.1 .

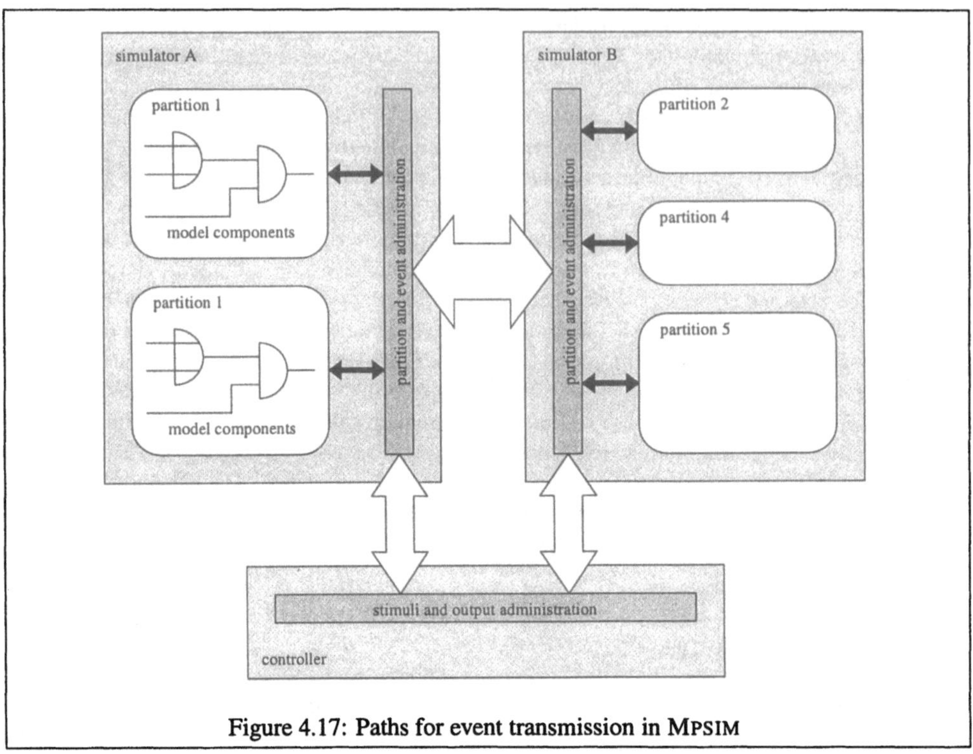

Figure 4.17: Paths for event transmission in MPSIM

Since the lifetime of model data is the whole simulation time, load balancing can only be done by data migration, not by placement of new entities. In this case, we also have to migrate model data like in the approach mentioned above. But here migration is only necessary in the case of a load balancing step, not when executing a single event.

In the following, we focus on regarding parts of model data as entities, since this is the approach, which can be combined with *Time Warp* best. For a successful load balancing strategy, entities must have a suitable size. The following paragraphs discuss several possibilities for entities. The logic simulator PSIM has two levels of hierarchy: single gates and partitions. It would be advantageous, if one of these abstractions could be used as entities.

Single gates: Single gates or signals (electrical lines) are natural, relatively independent entities of the circuit and could be used as entities. Since a single gate or signal represents only a very small part of the work of a simulator (typically 1/1000 to 1/100000), a lot of gates would have to be migrated in order to correct a certain practically relevant load imbalance. The migration of lots of single gates or signals would heavily disturb the initial partitioning of the circuit, which was optimized for low communication. Another kind of entities is needed, which preserves the quality of circuit partitions in respect of low communication between them [17; 15; 16]. The following approach offers this feature.

Multi partition approach: As already mentioned, using gates as entities is not an appropriate way of splitting the simulation work into subtasks. The next level of hierarchy within PSIM are circuit partitions. They are generated before simulation by the tool PARCOR and consist of several thousand gates and signals each. Unfortunately, a load balancing based on normal partitions cannot be done, because there exists only one partition per target in PSIM .

Based on the fact, that gates are too small and only one partition can be handled in PSIM , MP-SIM was developed. MPSIM has a similar simulation algorithm like PSIM and is able to simulate more than one partition on one simulator (see figure 4.17). **Partitions** represent the **workspaces** used in MPSIM . **Partitions now are entities,** which are migrated from one simulator to another in case of a load imbalance.

The number of partitions can be chosen freely when running the circuit partitioner PARCOR . It must not correspond with the number of simulators in the subsequent parallel simulation run. Thus, no modifications of the partitioning tool PARCOR are necessary, when used for PSIM or MPSIM . By choosing the number of partitions, static partition size (i.e. entity size) can be varied in a wide range. The partitions are initially placed on simulators for a statically well balanced start situation. The load balancing within MPSIM consists of **partitioning and assignment** of circuit partitions. The resulting **entity topology is a randomly connected graph**. PARCOR tries to minimize the edges of this graph, which represent the cut signals of the circuit. If we use the resulting partitions as entities for load balancing, cut costs and communication remains small even in the case of multiple migrations of entities/partitions.

4.4.2.2 Synchronization of Multiple Partitions

The synchronization mechanism of *Time Warp* itself needs a slight modification for MPSIM , because now several partitions reside in one target. They must be synchronized, too. Since partitions on a common target aren't simulated concurrently, synchronization between them can be conservative. *Time Warp* is used only for synchronizing the different targets running in different UNIX processes. There is no need for the *Time Warp* protocol within a target. Every partition holds its own event list and in general a simulator holds several partitions. By comparing the events with least execution times in all event lists residing on the target, the simulator can easily decide, which partition to simulate next. One single but complete simulation cycle for one time point in virtual time is done on that selected partition. This is only a slight modification to the normal simulation cycle in PSIM , which holds only one partition and its event list per simulator. More but smaller partitions residing on one simulators has the additional advantage, that rollbacks don't necessarily influence all model data on that simulator, but often only parts of it. The total number of rollbacks in MPSIM remains nearly the same as in the single partition case, although there are more and smaller partitions than in PSIM .

4.4.2.3 Targets

MPSIM 's target architecture is a **cluster of similar workstations**. For portability reasons, MP-SIM is based on normal, **heavy weight Unix processes**. Communication is implemented by the use of Unix sockets, which offer small latency. A future object oriented version will use the library PVM [9]. The physical media is Ethernet at the moment, but may be replaced by any other bus system. Small latency is very important for *Time Warp* . All partitions (i.e. entities, workspaces) placed on a certain processor (target) are maintained and simulated within the same Unix process. The processes are started at the beginning of the simulation. No additional processes have to be started (or ended) during simulation. The whole load balancing and migration of workspaces takes place within the initial processes.

4.4.2.4   A Special Load Index for *Time Warp*

The goal of load balancing for *Time Warp* is not homogeneous resource utilization (CPU time, memory). These numbers only have secondary importance. A *Time Warp* simulator is well balanced, if the number of rollbacks is minimal. A source for regular rollbacks is, if one simulator would be able to proceed in virtual time faster (or slower) than others. The goal for load balancing is to equalize the ability to proceed in virtual time.

As mentioned in section 4.4.2.2 , synchronization within one target is conservative. That means, that during normal execution there exists one common virtual time for all workspaces (partitions) residing on the same target (simulation process). Figure 4.18 shows the local virtual times (LVT_1 and LVT_2) of two simulators. Like in [5], our approach uses $LVTs$ for load balancing decisions. But compared to the simple standard deviation method in [5], a more sophisticated view on $LVTs$ is used.

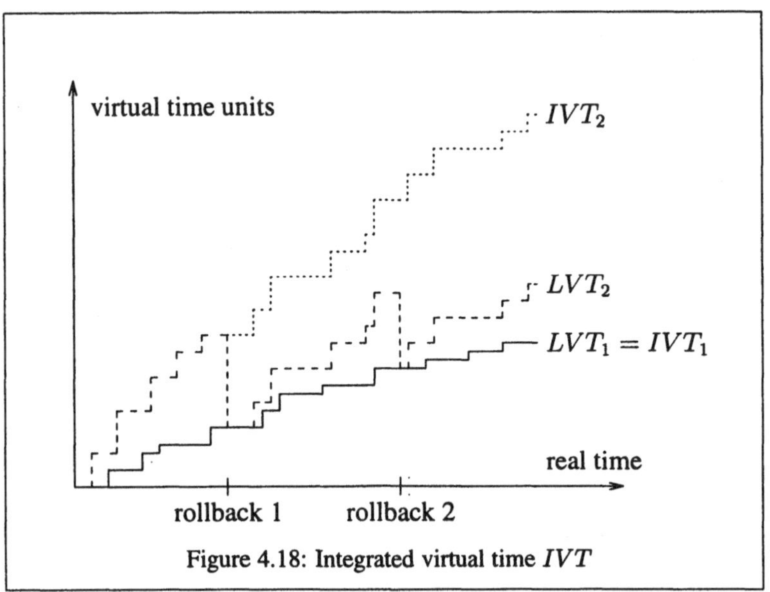

Figure 4.18: Integrated virtual time IVT

Underloaded simulators proceed faster in virtual time than others most of the time. Unfortunately, the reciprocal value of the rate of virtual time cannot be used as a load index directly, because of rollback influence. With or without rollbacks, long term virtual time rate is always equal on all targets (see LVT_1 and LVT_2 in figure 4.18). A rollback cleared rate of virtual time is needed, which means, that a virtual time point, which is calculated twice due to a rollback has to be taken into account twice, too. This calculation can be done on each simulator locally using the information, it has about event execution, rollbacks and local virtual time within the target, it is running on. We call this number virtual time progress (VTP). The following paragraphs show how it is calculated.

First, integrated virtual time IVT is introduced. IVT hides rollback influence and is measured in virtual time units. It is updated after every simulation cycle i:

$$IVT_i = \sum_{\nu \leq i} \left(\frac{\Delta T_\nu}{n_\nu} \right) \tag{4.4}$$

ΔT_ν is the increase in virtual time of simulation cycle ν (see section 4.4.1 on page 160) compared to cycle $\nu - 1$. n_ν is the number of partitions residing on this simulator (i.e. in this target) during simulation cycle ν. It prevents, that the same simulation cycle is counted multiply, once for each partition. If no rollback occurs, IVT is approximately equal to LVT (see LVT_1 and IVT_1 in figure 4.18). In case of a rollback and reexecution of a point in virtual time, IVT starts getting bigger than LVT (LVT_2 and IVT_2 in figure 4.18). LVT decreases, because it is determined by the event execution time of the actual event (straggler event, see section 4.4.1 on page 161). In contrast to LVT, IVT increases, because two simulation cycles are run for first execution and reexecution and both simulation cycles are included in the sum in equation 4.4.

The IVTs of subsequent simulation cycles are used to calculate VTP by:

$$VTP_i = \frac{IVT_i - IVT_{i-1}}{t_i - t_{i-1}} \tag{4.5}$$

IVT is calculated after each simulation cycle on each simulator asynchronously. The new value is **immediately sent** to the central controller which calculates the new VTP value of this simulator. The **central controller** has **complete information** about all simulators. **Index measurement** is not really periodic, but **implicit** depending on the activity of the simulated model. The load index VTP is **comprehensive** regarding **aggregation time** and totally **application related** regarding **aggregation space**. For stability reasons, the sequence of VTP values are additionally filtered in time. This smoothing prevents unnecessary migrations induced by short temporal effects in the simulation system.

Changes in VTP can have different reasons, like I/O speed, CPU speed, CPU load, event generation, swapping, ... For example, if the difference in virtual time between subsequent simulation cycles gets bigger in average due to model behavior, the difference between subsequent IVTs and therefore VTP increases. This is correct, because, this simulator advances faster in virtual time. In another situation, there could be more events to simulate during one simulation cycle. This would increase the real time difference in the denominator of equation 4.5. VTP will decrease, which is again correct, because the simulator gets slower, when there are more events to execute.

Load balance can be disturbed by a lot of internal or external influences. A general load balancing tool, which is not incorporated in the application only can observe these influences and guess, how they change the behavior of the application program. This is a source for non-optimal decisions of the load balancing algorithm. The advantage of the reciprocal VTP as load index is its strong **application adaptivity**. The way, the load index itself is calculated is **fixed**. Nevertheless, it has a high automatic application adaptivity by the usage of strongly application related numbers for VTP calculation. For *Time Warp* it makes no difference, what the reason for differing VTPs in targets is (application internal or external). For a good load distribution in *Time Warp* it is only important, that VTPs in all targets are as similar as possible, because this is the main supposition for a small number of rollbacks. Therefore the reciprocal value of VTP is a very good **single dimensional load index** in a *Time Warp* simulation environment. By usage of VTP as a single load index, **monitoring** is completely done within the application. No system information is needed.

4.4.2.5 Load Balancing Strategy

The load balancing in MPSIM consists in an exchange of a partition between the fastest and the slowest simulator. Only **two targets participate** in a balancing step. The migration of a partition is initiated by the central controller and stops the sending and receiving simulator for a certain time. A successful load balancing strategy may not neglect this time. The performance gain (increase in VTP) after migration must be big enough to justify the momentary slow down. Two parameters are important in this context:

- t_{rec}: the time, a migration lasts and
- VTP_p: predicted VTP of the sending simulator after the migration.

t_{rec} depends on the amount of data, which has to be transmitted and can be approximated by a constant during a single simulation run, because all partitions nearly have the same size. By taking t_{rec} into consideration, MPSIM has a **high migration cost sensitivity**. After migration a simulator with previously n partitions residing on it has a predicted VTP_p of:

$$VTP_p = VTP \frac{n}{n-1} \tag{4.6}$$

In the case, that another simulator is the slowest one after migration, a more complex formula (not mentioned here) is used for predicting VTP_p. Knowing VTP_p and t_{rec}, a break even time t_{BE} can be calculated (see figure 4.19) by the central controller of the simulation system. After a migration it lasts t_{BE} until the sending simulator has reached the same LVT like without migration.

$$t_{BE} = \frac{VTP_p \cdot t_{rec}}{VTP_p - VTP} \tag{4.7}$$

A migration is only useful, if t_{BE} is small enough. The break even point should be reached before the load situation changes again. For this reason, a maximum $t_{BE_{max}}$ is introduced. The decision, whether a migration is initiated or not is done threshold–based on this number and on the difference in VTP of the fastest and slowest simulator. After transformation, we get the following **analytic condition** for initiating a migration at runtime (**dynamic decision process**):

$$VTP_p > \frac{t_{BE_{max}}}{t_{BE_{max}} - t_{rec}} VTP \tag{4.8}$$

$t_{BE_{max}}$ can be **adjusted at simulation startup**. On a system with low load fluctuations, a high value may be used and on a system with a fast dynamic behavior, a low value should be used. It is possible to control flexibility of the simulation system by this single number. In the current

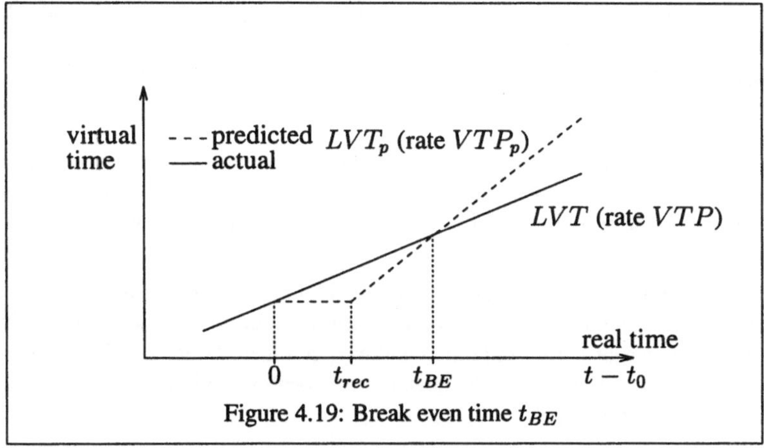

Figure 4.19: Break even time t_{BE}

version of MPSIM experiments must show the best value for $t_{BE_{max}}$. This value may be used for all future runs, because in most systems the dynamic behavior remains basically the same. An appropriate value of t_{rec} and the smoothening of VTP values (see section 4.4.2.4 on page 165) are a **weak internal stability control**.

The direct relation of that condition (equation 4.8) to numbers given by the application shows, that this **strategy** is strongly **application oriented**, not system oriented.

4.4.2.6   Migration of a Partition

Once, the decision is taken to do a load balancing step, a circuit partition (i.e. workspace) has to be migrated from one target to another. **Migration set is limited** to a single partition. **Transfer policy is preemptive**, because lifetime of a circuit partition is the whole simulation run and therefore it makes no sense to wait for completion of the simulation of a single partition. Furthermore, **transfer policy is complete**, i.e. a partition is transfered completely in a single step. The transfer is implemented fully **within the application** and can be done locally by the two participating simulators after they have received the corresponding order from the central controller and finished the actual simulation cycle (see section 4.4.1 on page 160), which is only a **short initiation delay** compared to migration time. There is no limitation, which simulators may exchange a partition, so **transfer space is system wide**. All other simulators continue simulating. The **migration** of a partition between two other simulators is **transparent** to all others and to the user. Due to the integration of the load balancing and migration into the *Time Warp* application, **no source code transparency** is possible. The data, which has to be transmitted binary and inserted in the structures of the destination simulator consists of the following:

- **Circuit model:** the data structures representing the gates and signal lines of the partition, which is migrated. This includes the static circuit topology and the states of the circuit components at actual LVT.
- **Event list:** All pending events for the migrated partition are transferred. They will be simulated on the new target.

- **Saved states:** The state copies for the migrated partition have to be transmitted. They will be needed on the new target, if a rollback occurs.
- **Administrative data:** The simulator itself holds some variables about the circuit, too (e.g. LVT). They are sent to the new target simulator.

In MPSIM pending events and saved states are both stored in a common event list [3]. By migrating the event list, events and states are transmitted commonly. No resources from other nodes other than the one hosting the partition are needed before or after migration. There are **no residual dependencies**.

After successful migration of a partition, all other simulators are informed about its new host target. Although new events are sent to the correct new location from now on, there may exist some events, which were sent before migration and have not there target yet. They will arrive at the old target and are just forwarded to the new, correct location without using resources worth mentioning on the forwarding target.

4.4.3 Classification Tables

The following tables show, how the load balancing approach within MPSIM can be classified according to the global classification scheme.

References

[1] H. BAUER, C. SPORRER and H. T. KRODEL. On Distributed Logic Simulation Using Time Warp. In *IFIP Int. Conf. on Very Large Scale Integration (VLSI)*, pages 127–136, Edinburgh, August 1991.

[2] H. BAUER and C. SPORRER. Distributed Logic Simulation and an Approach to Asynchronous GVT-Calculation. In *ACM/SCS/IEEE WS on Parallel and Distributed Simulation (PADS)*, pages 205–209, 1992.

[3] H. BAUER and C. SPORRER. Reducing Rollback Overhead in Time-Warp Based Distributed Simulation with Optimized Incremental State Saving. In *SCS/IEEE Annual Simulation Symposium (ASS)*, 1993.

[4] H. BAUER. *Verteilte diskrete Simulation komplexer Systeme*. Ph.d. thesis, Institute of Electronic Design Automation, Technical University of Munich, 1994.

[5] C. BURDORF and J. MARTI. Load Balancing Strategies for Time Warp on Multi-User Workstations. *The Computer Journal, British Computer Society*, 36(2):168–176, 1993.

[6] J. CLEARY, F. GOMES, B. UNGER, X. ZHONGE and R. THUDT. Cost of State Saving and Rollback. In *ACM/SCS/IEEE WS on Parallel and Distributed Simulation (PADS)*, pages 94–101, 1994.

[7] L. M. D'SOUZA, X. FAN and P. A. WILSEY. pGVT: An Algorithm for Accurate GVT Estimation. In *ACM/SCS/IEEE WS on Parallel and Distributed Simulation (PADS)*, pages 102–109, 1994.

[8] J. FLEISCHMANN and P. A. WILSEY. Comparative Analysis of Periodic State Saving Techniques in Time Warp Simulators. In *ACM/SCS/IEEE WS on Parallel and Distributed Simulation (PADS)*, pages 50–58, Lake Placid, New York, June 14-16 1995.

[9] A. GEIST, A. BEGUELIN, J. DONGARRA, W. JIANG, R. MANCHEK and V. SUNDERAM. *PVM: Parallel Virtual Machine. A Users' Guide and Tutorial for Networked Parallel Computing*. Scientific and Engineering Computation Series. MIT Press, Cambridge, MA, 1994.

[10] D. R. JEFFERSON. Virtual Time. *ACM Trans. on Programming Languages and Systems*, 7(3):404–425, 1985.

[11] B. KANNIKESWARAN, R. RADHAKRISHNAN, P. FREY, P. ALEXANDER and P. A. WILSEY. Formal Specification and Verification of the pGVT Algorithm. In *Proc. Formal Methods Europe*, Oxford, Mar 1996.

[12] A. PALANISWAMY and P. WILSEY. An Analytical Comparison of Periodic Checkpointing and Incremental State Saving. In *ACM/SCS/IEEE WS on Parallel and Distributed Simulation (PADS)*, pages 127–134, San Diego, 1993.

[13] M. RUHWANDL. Verteilte Logiksimulation mit vervielfachter Partitionsanzahl. Master's thesis, Institute of Electronic Design Automation, Technical University of Munich, 1992.

I. General

Objectives

domain specific	yes
intent	application adaptive
function	assignment & partitioning

Integration Level

monitoring	application
strategy	application
mechanisms	application
compiler support	none

Structure

problem structure	location dependent
subproblems	data access
workspaces	data
targets	processes

III. Load Model

Load Index Properties

index dimension	single-dimensional
index type	application defined
aggregation time	comprehensive
aggregation space	application related

Load Value Composition

index combination	—
weight selection	—

Model Usage Policies

index measurement	implicit
index propagation	after measurement
model adaptivity	fixed

II. Strategy

System Model

Model Flavour	analytic
Target Topology	NOW
Entity Topology	randomly connected graph

Transfer Model

Transfer Space	system wide
Transfer Policy	preemptive

Information Exchange

Information Space	central
Information Scope	complete

Coordination

Decision Structure	centralized
Decision Mode	competitive
Participation	partial

Algorithm

Decision process	dynamic
Initiation	threshold–based
Adaptivity	adjustable (at startup)
Cost Sensitivity	high
Stability Control	weak

IV. Migration Mechanism

Migrant	data
Migration Set Size	limited
Heterogeneity	no
Migration Initiation	delayed
Pre Transfer Media	none
Compression	no
Transfer Policy	complete
Residual Deps.	none
Migration Transp.	yes
Source Code Transp.	no

Table 4.7: Classification Tables

[14] R. SCHLAGENHAFT, M. K. RUHWANDL, C. SPORRER and H. BAUER. Dynamic Load Balancing of a Multi-Cluster Simulator on a Network of Workstations. In *ACM/SCS/IEEE WS on Parallel and Distributed Simulation (PADS)*, pages 175–180, Lake Placid, New York, June 14-16 1995.

[15] C. SPORRER and H. BAUER. Partitioning VLSI-Circuits for Distributed Logic Simulation. In *SCS European Simulation Multiconference (ESM)*, pages 409–413, 1992.

[16] C. SPORRER and H. BAUER. Corolla Partitioning for Distributed Logic Simulation of VLSI-Circuits. In *ACM/SCS/IEEE WS on Parallel and Distributed Simulation (PADS)*, pages 85–92, 1993.

[17] C. SPORRER. *Verfahren zur Schaltungspartitionierung für die parallele Logiksimulation.* Ph.d. thesis, Technical University of Munich, 1994.

5 Summary

There are many possible load distribution techniques for parallel applications. This book surveys the present approaches to load distribution, emphasizing *dynamic* load distribution. It shows classifications and describes several examples in detail. The examples range from a general process migration concept to domain-specific and application-integrated implementations.

Chapter 2 introduced classifications of load distribution. The classifications are rather pragmatic than theoretical. We did not discuss every theoretically possible solution but considered practically important techniques. We have tried to be independent of a specific implementation or integration level. Therefore, the classifications use the general terms *entities* and *targets* for load distribution. The authors listed in the beginning of the book, working in rather different domains of parallel computing, have not only applied the classifications, but they have also contributed to the classifications. Therefore, we are confident that the classifications cover almost all aspects of dynamic load distribution for parallel applications.

The first classification is a *general classification* regarding the domain and integration level of load distribution. The most important attribute concerning the structure of load distribution, namely whether subproblems are location dependent or not, were discussed and illustrated by examples. Load distribution *strategies* are classified in the subsequent section. We have discussed possible system models, possible cooperation between load distribution instances and possible algorithms. Next, *load models*, representing the information on which load distribution strategies rely, are classified. We showed that load models may range from unique values such as queuelength to measures that use complex formulas. Furthermore, there are several possibilities how load information is propagated. A further classification considered *migration mechanisms* because migration of entities such as processes and application objects is a complex task. These classifications have been applied to several examples in Section 2.5.

Chapter 3 presented systems supporting load distribution. Such systems aid developers of parallel applications using load distribution techniques. *CoCheck* (Section 3.1) can be used to transparently migrate processes of PVM applications. Section 3.2 described load distribution strategies that can be integrated into a distributed thread kernel. These strategies were classified according to the strategy classification scheme in Chapter 2. *Dynasty* (Section 3.3) is an economic-based concept for load distribution in workstation networks. In Section 3.4, we described ALDY, a library supporting application-integrated load distribution by dynamically assigning and migrating application objects.

Chapter 4 contains load distribution techniques that have been developed for specific application domains. Section 4.1 described a middleware-based architecture for load distribution. Load distribution strategy and mechanisms are transparently integrated into the runtime system of the underlying IDL-environment. Sections 4.2 and 4.3 showed two applications out of different domains that use application specific information for dynamic assignment of subproblems. Whereas Section 4.2 dealt with parallel automated theorem proving, Section 4.3 described the generation of test patterns for digital integrated circuits. In both applications, redundant work can be prevented by preferring "important" subproblems for dynamic assignment. Section 4.4 showed an example of a *time warp* application that uses migration of circuit partitions for load distribution.

All sections in Chapter 3 and 4 applied the classifications in Chapter 2 to the given approach. Of course, some classification attributes were not fixed in some cases, since several of the presented solutions are generic concepts.

Our own experiences with the classifications were that they are not only useful to obtain a survey of load distribution, but aid in the understanding of unfamiliar work. Approaches that at first sight looked rather dissimilar due to the use of different terms in their descriptions turned out to be of the same kind.

The classifications schemes and the subsequent examples may serve novices as a survey over the state of the art of dynamic load distribution for parallel programs. Experts in this area will hopefully make use of the general terminology that is not based on a specific implementation or integration level.

About the Authors

Martin Backschat: Section 3.3

> Institut für Informatik, Technische Universität München, Munich, Germany
> Email: backscha@informatik.tu-muenchen.de
> WWW: http://wwwzenger.informatik.tu-muenchen.de/persons/backscha.html

Stefan Bischof: Section 2.2

> Institut für Informatik, Technische Universität München, Munich, Germany
> Email: bischof@informatik.tu-muenchen.de
> WWW: http://wwwmayr.informatik.tu-muenchen.de/personen/bischof

Andreas Ganz: Section 4.3

> Institut für Schaltungstechnik, Technische Universität München, Munich, Germany
> Email: asg@regent.e-technik.tu-muenchen.de
> WWW: http://www.regent.e-technik.tu-muenchen.de/people/asg.html

Thomas Erlebach: Section 2.2

> Institut für Informatik, Technische Universität München, Munich, Germany
> Email: erlebach@informatik.tu-muenchen.de
> WWW: http://wwwmayr.informatik.tu-muenchen.de/personen/erlebach

Marc Fuchs: Section 4.2

> Institut für Informatik, Technische Universität München, Munich, Germany
> Email: fuchsm@informatik.tu-muenchen.de
> WWW: http://wwwjessen.informatik.tu-muenchen.de/personen/fuchsm.html

Claudia Gold: Section 3.4

> Institut für Informatik, Technische Universität München, Munich, Germany
> Email: gold@informatik.tu-muenchen.de
> WWW: http://wwwpaul.informatik.tu-muenchen.de/personen/gold.html

Alexander Pfaffinger: Section 3.3

> Institut für Informatik, Technische Universität München, Munich, Germany
> Email: pfaffina@informatik.tu-muenchen.de
> WWW: http://wwwzenger.informatik.tu-muenchen.de/persons/pfaffina.html

Markus Pizka: Section 3.2

> Institut für Informatik, Technische Universität München, Munich, Germany
> Email: pizka@informatik.tu-muenchen.de
> WWW: http://wwwspies.informatik.tu-muenchen.de/personen/pizka/

Niels Reimer: Section 3.2

> Institut für Informatik, Technische Universität München, Munich, Germany
> Email: reimer@informatik.tu-muenchen.de
> WWW: http://wwwspies.informatik.tu-muenchen.de/personen/reimer/

Christian Röder: Section 2.3

> Institut für Informatik, Technische Universität München, Munich, Germany

  Email: roeder@informatik.tu-muenchen.de
  WWW: http://wwwbode.informatik.tu-muenchen.de/ roeder

Björn Schiemann: Section 4.1
  Siemens AG, ZT SW 2, Munich, Germany
  Email: Bjoern.Schiemann@zfe.siemens.de

Rolf Schlagenhaft: Section 4.4
  Institut für Schaltungstechnik, Technische Universität München, Munich, Germany
  Email: ros@regent.e-technik.tu-muenchen.de
  WWW: http://www.regent.e-technik.tu-muenchen.de/people/ros.html

Dr. Thomas Schnekenburger: Editor, Sections 2.1, 2.5, and 3.4
  Institut für Informatik, Technische Universität München, Munich, Germany
  Email: schneken@informatik.tu-muenchen.de
  WWW: http://wwwpaul.informatik.tu-muenchen.de/personen/schneken.html

Dr. Georg Stellner: Editor, Sections 2.4 and 3.1
  Institut für Informatik, Technische Universität München, Munich, Germany
  Email: stellner@informatik.tu-muenchen.de
  WWW: http://wwwbode.informatik.tu-muenchen.de/ stellner

Andreas Wolf: Section 4.2
  Institut für Informatik, Technische Universität München, Munich, Germany
  Email: wolfa@informatik.tu-muenchen.de
  WWW: http://wwwjessen.informatik.tu-muenchen.de/personen/wolfa.html

Heidtmann
Zuverlässigkeits-bewertung technischer Systeme

Modelle für Zuverlässigkeits-strukturen und ihre analytische Auswertung

Dieses Buch ist eine Einführung in das Gebiet der Zuverlässigkeit technischer Systeme. Dargestellt werden gesellschaftliche und praktische Aspekte sowie die grundlegenden Begriffe und Modelle der Zuverlässigkeit und insbesondere die Verfahren zur Zuverlässigkeitsbewertung. Zahlreiche anschauliche Beispiele erleichtern das inhaltliche Verständnis, zeigen die Möglichkeiten und Grenzen der praktischen Anwendung und demonstrieren die durch neue Forschungsergebnisse erzielten Fortschritte.

Das vorliegende Buch führt alle, die sich aus gesellschaftlicher oder technischer Sicht für die Zuverlässigkeit der Technik interessieren, in dieses wichtige Gebiet ein. Darüber hinaus soll es Studierenden mathematischer, naturwissenschaftlicher und technischer Fachrichtungen zum Selbststudium oder als Begleitmaterial zu Vorlesungen oder Schulungen dienen und nicht zuletzt den in der Industrie, in Versorgungsunternehmen, Behörden, wissenschaftlichen und technischen Institutionen usw. Tätigen bei Zuverlässigkeitsfragen und -problemen weiterhelfen.

Von Dr.
Klaus Heidtmann
Universität Hamburg

1997. 220 Seiten mit 39 Bildern und 7 Tabellen. 16,2 x 23,5 cm. Kart. DM 54,80 ÖS 400,– / SFr 49,– ISBN 3-8154-2306-6

(TEUBNER-TEXTE zur Informatik, Bd. 21)

B. G. Teubner Stuttgart · Leipzig